Development through Digitization:
Addressing the LDC Challenge

Darrell E. Owen

Copyright © 2017 Darrell E. Owen

All rights reserved

ISBN-10: 1978279124
ISBN-13: 9781978279124

Dedication

This book is dedicated to those who in the past have been, those who currently are, and those who in the future will become, engaged in expanding affordable Internet access and delivering socioeconomically impacting services across the Internet—specifically delivering services to those living in the rural areas of the Least Development Countries (LDCs).

About the Author

Darrell Owen has worked in the international development arena for nearly 25 years (1993-2017). He worked as an employee for the U.S. Agency for International Development (USAID) for 6 years. He then consulted for 18 years through over a dozen for-profit and non-profit firms that held USAID contracts and grants. Prior to joining USAID, Darrell worked at the Bonneville Power Administration (BPA) for 18 years, beginning in 1970 after completing 3 years military service in the U.S. Army.

He first became involved with information technologies (IT) in the late 1960s while in the U.S. Army. His IT-related work continued throughout his working experience, with a concentrated focus during the last decade on expanding affordable voice and Internet to those living in rural communities. He holds a Bachelor's Degree in Business Management from Portland State University (Portland, Oregon), and a Master's Degree in Telecommunications from the George Washington University (Washington, DC).

Darrell has been actively engaged in a wide range of ICT-related programs and projects. This has included designing and launching new initiatives, leading countrywide assessments, designing country-level programs, and managing and executing on-the-ground projects.

Recently, his work focused on increasing affordable Internet access as a foundational component for expanding socioeconomic opportunities, especially to the rural populations who are most often ignored.

Acknowledgements
...and Thanks

This book reflects a compilation of my work experiences in partnership with many others. It is also the result of discussions with, and encouragement from, others from countries across our planet with whom I had the privilege of working. Following are just a few of those who had a major impact on my work.

Bernie Mazer—a USAID colleague who supported my transition from working inside USAID to moving into the field of consulting. A good friend with whom I learned a lot and enjoyed working with on many trips—including hours spent perusing bookstores in London as we traveled to, or returned from our overseas travels.

Bob Otto—Bob and I first met in Jamaica when he was VP for Carana Corporation. We were engaged in a project to assist the country in writing new legislation to liberalize the local mobile market. Later, we worked together in support of a telecom-liberalization effort for the OECD countries. When Bob formed his own business in 2010, Integra LLC, his firm played a critical role in supporting USAID's Global Broadband and Innovation (GBI) program. More recently, Integra LLC supports ICT projects for the Millennium Challenge Corporation (MCC).

David Mendoza—a colleague during my early consulting years—working jointly on a range of ICT-related projects in Armenia, with a foray into Eritrea. I learned much from his take-charge orientation, his tackling of challenging projects with potentially big impact, and his building relationships with local on-the-ground partners.

David Townsend—a consulting colleague that played a pivotal role in several initiatives, starting with USAID's Last Mile Initiative (LMI) in Vietnam during the early-mid 2000s. In Vietnam, he set the standard for operationalizing the country's newly authorized Vietnam Telecommunications Fund (VTF). He later worked on a range of initiatives through the Global Broadband & Innovation (GBI) Program, launching new and refining existing Universal Services Funds (USFs).

Bernd Nordhausen—a colleague who I worked closely with in Vietnam through Intel's partnership with USAID under the Last Mile Initiative (LMI). Bernd had the lead in deploying two WiMAX networks in the Lao Cai Province—one in the capital and the other in the rural village of TaVan.

David Lyman—served a critical role in rapidly-deployable, low cost, wireless solutions, including providing wireless voice and Internet solutions. Dave was engaged in introducing the Internet in Africa through the Leland Initiative (LI). His work extended to supporting the LMI in both Vietnam and Mongolia, with a focus on Voice over IP (VoIP), and later via small cell solutions.

Steve Schmida—a colleague and friend with whom I had the opportunity to work and learn from on several initiatives, including USAID's the Last Mile Initiative (LMI) in Sri Lanka and Vietnam. More recently Steve brought me into playing a small role on two research initiatives with USAID and DIAL through his firm, SSG-Advisors.

Andrew Reynolds—a colleague who I met when he worked at the U.S. State Department. He had faith and conviction in the work I was engaged in. Between 2012-2015 he tagged me to represent our country and make a series of presentations—where I pulled together panels for four of the annual meetings of the UN's Commission on Science and Technology for Development (UNCSTD/CSTD) in Geneva.

Joe Duncan—a colleague and friend at USAID who reached out to me around the 2008 timeframe. His request was straightforward. He wanted my help to design a new rural-focused connectivity program—not a program designed from Washington DC, rather, one that was based upon on-the-ground, in-country successes. This was an area where I had been engaged for nearly 10 years, and from that experience we designed and launched the Global Broadband and Innovation (GBI) program.

Dave Ferguson—I first met Dave when he was exploring possible engagement in the international development space in the mid-2000s. Prior to this time, he had a career with AT&T. We teamed up on two overseas trips to the countries of Georgia and South Sudan. Later, he joined the USAID workforce, where he helped in the early design of the GBI program. He added the "innovation" to the GBI Program.

Rita Owen—my wife and best friend for nearly 55 years. She led the way in getting us to move to Washington DC to join USAID in the mid-1990s, knowing we could both make a difference in something we believed in. After I took an early retirement from USAID to start consulting, she put up with my frequent absences when I spent weeks, even months, on the road pursuing my passion.

Table of Contents

Dedication

About the Author

Acknowledgements and Thanks

Table of Contents

Introduction

1.	First Bookend -- Wireless Village: First Mile First	1
2.	Dynamics of 2000-2015	13
3.	Future Direction: 2015-2030	39
4.	LDCs: The Ultimate Challenge	53
5.	"Digitization" and Other ICT Ecosystems	63
6.	Second Bookend -- Development through Digitization Model	89
7.	Sharpening the Focus for the Next 15 Years	123

Concluding Thoughts

Acronyms

Introduction

The foundation for this book, "Development through Digitization: Addressing the LDC Challenge," is passion—passion that has been the primary driver for my international engagements. This book seeks to share key observations, experiences, and lessons-learned. From this base, I also project forward some of the anticipated future dynamics that will dominate the next 15 years, as well as thoughts for advancing the agenda.

This experience-based focus is augmented with related research and references. While the book has a specific focus on the Least Developed Countries (LDCs), the contents are equally applicable for countries in the lower and middle tiers of developing economies.

> "The Digital Divide continues to be the Development Divide"
>
> Irina Bokova,
> Director General
> UNESCO

The LDC *Challenge* reflected in the title has been with us for decades. It was perhaps best stated several years back by Irina Bokova, the Director General of UNESCO.

Orientation of the Book

This book was written with a specific orientation that includes the following four threads:

1. Establishing an overview of the ICT-related initiatives undertaken during the 2000-2015 timeframe of the Millennium Development Goals (MDGS) developed by the United Nations (UN). It also includes more recent goals established within the context of the 2015-2030 timeframe of the Sustainable Development Goals (SDGs);

2. Capturing and putting forward key elements of my engagements in this space, predominantly during the timeframe covered by the MDGs, and the first couple years of the SDGs;

3. Putting forward the rationale, as well as a model, for analysis and action that supports future development through a comprehensive Development through Digitization (DtD) model; and

4. Focusing specific attention on the need for the international community to expand its support to achieve the socioeconomic benefits of leveraging digitization by those living in the Least Developed Countries (LDCs). That is, moving from discussion to action.

Structure of the Book

The structure of this book is designed around two chapters (Chapter 1 and Chapter 6). These two chapters serve as bookends.

First Bookend: Wireless Village - The foundational theme was established in late 1998 when I was nearing the completion of my Master's Degree in Telecommunications from the George Washington University (GWU). At that time, I completed a research project titled, *"Wireless Village: First-Mile-First."* Paradoxically, while the focus of my research looked forward, the fundamental elements from this research dated back to the early 1900s. It was a time when the explosion of thousands of independent local community networks drove the rural build-out of telephony in the United States. Hence the title. This research paper serves as historical precedence for expanding rural Internet in developing countries today.

Second Bookend: Development through Digitization (DtD) Model – The second bookend is based on a paper I initially wrote in 2015—where the rural connectivity need is not so much for voice services, but rather for expanding and leveraging the Internet. The DtD model was developed based on my personal on-the-ground engagements in developing country settings, including several Least Developed Countries (LDCs). This is where the ultimate challenge to achieve this leverage remains unmet, and where the gap continues to grow. This second bookend expands the focus: from Connectivity to Digitization.

MDGs and SDGs - Between these two bookends, attention is placed on the information, communications and technology (ICT) dynamics that have been, and currently are, taking place within the international development community. This is divided into two 15-year Chapters—the Millennium Development Goals (MDGs) covering the timeframe of 2000-2015, and the launch and initial years of the Sustainable Development Goals (SDGs) covering the timeframe of 2015-2030.

LDC Challenge - The second topic covered between these two bookends further details the characteristics of the Least Developed Countries (LDCs). The key data that is exhibited goes beyond the most current adoption data, and provides country profiles of the population living in these countries. Here the challenge quickly surfaces with regards to the percent of population living in extreme poverty, as well as the percent of population living in rural areas. This population is a strategic focus of both the MDGs and SDGs. Combined with other variables, these populations are the fundamental challenge associated with expanding the reach of affordable Internet.

Emerging Theme of Digitization - The third topic covered between the two bookends is the growing recognition that attention needs to be placed well beyond simply expanding affordable Internet access. To achieve the sought-after socioeconomic impact for those living in rural villages, it takes much more than just adding connectivity. It requires a fully integrated ecosystem of essential elements. This emphasis is captured through the term, "Digitization," though some refer to this concept as "Digitalization." The chapter is devoted to an overview of source materials that present the value to be gained from this expanded orientation called Digitization.

Sharpening the Future Focus - The book concludes with a focus on the next 13 years—the remaining years of the SDGs. In this chapter, the topic builds off the current state and sharpens the future agenda. This is where there is the opportunity to rethink—ideally where "Digitization" becomes the norm as we move forward.

...experiences and observations from the author

In the fall of 1999, I retired from USAID. After retiring, I established relationships with several firms that had contracts with USAID on a range of ICT-related projects. One of my consulting engagements was supporting the Program Technology Transfer (PTT) initiative. The PTT provided consultations to USAID/Washington and USAID's overseas Missions. It also supported the White House's Internet for Economic Development (IED) Initiative. In addition to this work, I consulted through several firms for stand-alone USAID funded projects where there was an imbedded ICT component, or the project had a specific ICT focus.

During my consulting in the international technology arena, I did ICT-related assessments in an estimated 20+ countries, with follow-up implementing engagements in several countries (Armenia, Eritrea, Mongolia, Vietnam, Sri Lanka, Kenya, Jamaica, Indonesia and others). In most of these countries my focus was on advancing their connectivity, leveraging ICTs across multiple sectors, and expanding the local ICT sector.

In the 2015 timeframe I developed a thought piece that I referred to as "Digitization for Development" (DfD). This paper captured lessons coming from my work and pulled them into a single integrated framework—a model that included a wide-range of variables ultimately needed for achieving success. In writing this book, I retitled, refined, and expanded this model. In this book I now refer to it as "Development through Digitization" (DtD)—shifting the primary focus to Development, with Digitization repositioned as having a supporting role. I also sought to place a priority on the Least Developed Countries (LDCs).

The rationale for writing this book is simple: to pass on lessons-learned to others. It is not put forward as "the way to do it," but rather as illustrative examples of what has worked in some of these countries, as well as what has not worked.

My personal experiences and observations are captured and presented throughout this book within the context of the addressed topics. They are in the form of text boxes labeled, *"...experiences and observations from the author."*

CHAPTER 1

- First Bookend -
Wireless Village:
First-Mile-First

This first chapter is an article-length version of a graduate research paper I completed in September 1998 on the topic, "Wireless Village: First-Mile-First." The research paper represents the first bookend. The second bookend, "Development through Digitization," is captured in Chapter 6.

Between these intervening years, the dominant focus with regards to connectivity has shifted from expanding telephony to expanding affordable Internet—a technology that offers greater socioeconomic potential, but also significantly greater challenges.

Wireless Village: First-Mile-First—The following is the abstract from the original research paper:

> *"As the world moves toward globalization, telecommunications increasingly becomes one of the primary topics demanding attention. In recent years the Internet has become the center of this focus--not simply for sharing information, but as the backbone facilitating growth in transnational commerce. Yet in the shadows of the Internet, most of the world's population is without simple access to a telephone. This is especially acute in the rural populations of developing countries where one telephone may exist for every 1,000-3,000 individuals; where access to a telephone may be a day away, perhaps more.*
>
> *"This paper focuses on rural connectivity--the "last mile." Specifically, it proposes that the last mile receive priority attention; that local access be viewed as the "first mile." While expanding worldwide communications is essential, ultimately communications is a local issue--requiring affordable local access. This local focus is not a counterpoise to the broader national and international agenda, but rather a vital complementary component.*
>
> *This paper proposes an approach that holds promise for accelerating the rate of build-out by building in; by first meeting local requirements. The paper extracts valuable lessons from the U.S. experience in the late 1800s/early 1900s where there was a virtual explosion of telephony expansion in rural areas. This was brought about by literally thousands of local independent telecommunication companies--where expansion*

took place in a massively parallel manner at the local community level. The paper leverages the current trend toward market liberalization aimed at reducing long-standing barriers that have restricted access to technology and limited competition. The wireless technologies and solar power are critical enabling technologies that make it feasible for serving even the lowest income and lowest density rural populations. While there are an overwhelming number of obstacles to overcome, the paper concludes with an action outline aimed at servicing the currently unreached in an economically-sustainable manner, and in doing so, set the stage for even greater connectivity to rural areas."

The original research paper was 72 pages in length, and due to its size, it is simply too large to include within this book. However, in May of 1999, I created a shorter, 7-page overview. This shorter version is the source material for this first chapter.

Those wishing to have access to the original research paper can contact the author at darrell_owen@msn.com, and I will forward to you an electronic copy (PDF) via e-mail, free-of-charge.

...experiences and observations from the author

During the timeframe of getting my Master's degree in Telecom from George Washington University and writing this research paper, I worked at the U.S. Agency for International Development (USAID). At the time, I managed USAID's "Missions Support" unit that included providing ICT-related support to both internal operations as well as programmatic engagements, both domestic and overseas.

Within a year of completing the Wireless Village paper and taking an early retirement, I launched into my post-retirement career as an independent international telecommunications consultant. My initial consulting engagements supported USAID's ICT overseas programs, but to begin with only a few projects were in expanding connectivity. Later, expanding connectivity became a more dominant theme as well as the majority of my consulting work.

Wireless Village: First Mile First

May 11, 1999

With the dominant focus today on the Internet, high-speed fiber, and mega-mergers within the telecommunications industry, it's easy to overlook the fact that most people living on this planet today do not have simple access to a telephone. With 80 percent of the world's population living in developing countries, this phone-less situation is not limited to a few isolated locations.[1] Rather, it is the rule. This is especially acute in rural areas of developing countries—where on average 70 percent of their people live and where the number of phones may be on the order of one phone per 100, even one per 1,000 inhabitants; in some locations even less.[2]

Fortunately, there has been an increasing focus on this situation over the past decade or so, with accelerated focus in more recent years. In the early 1980s, Sir Donald Maitland chaired the Independent Commission for Worldwide Telecommunications Development Commission. It was from this effort that "The Missing Link" report was issued in December of 1984—highlighting this situation and triggered much subsequent action. More recently the World Trade Organization (WTO) has played a critical role in bringing countries together to negotiate the Agreement on Telecommunication Services signed by 69 countries in 1997. Unfortunately, many of those countries most in need didn't sign on. However, a growing number of developing countries are putting new telecommunications policies in place that move their domestic markets into a more liberal position—allowing multiple players and competition. This is typically supported with regulatory reforms, establishing separate regulatory bodies, putting in place new administrative provisions for settling disputes, etc. Many developing countries are beginning to focus specifically on extending the reach of their current public networks into rural areas.

While these are all good, the question is whether this is good enough. Or more to the point, "will these efforts have an impact quick enough?" as Thomas Friedman points out in his book, "The Lexus and the Olive Tree," Friedman, Thomas L., published by Farrar, Straus & Giroux. New York, NY. March 1999.[3] Today there is the *fast world* and the *slow world*. The longer it takes, the further behind these slow countries will be when they do get connected. "Leapfrogging" seems a distant illusion.

---

[1] Kayani, Rogati and Dymond, Andrew. Options for Rural Telecommunications Development (World

[2] National Telephone Cooperative Association. A Solution for Serving Rural Areas.
http://www.ntca.org/intl/solution/index.html

[3] Friedman, Thomas L., "The Lexus and the Olive Tree," Farrar, Straus & Giroux. New York, NY. March 1999

A recent report by the World Bank reflects that even under optimal circumstances, gaining any meaningful teledensity in rural areas takes on the order of 10 years, typically 20 or more—a period during which the fast world is even further down the road. Clearly there must be a better approach.

Back to the Future

Interestingly, perhaps the most promising approach for accelerating access to rural areas is over 100 years old. Unfortunately, the experience was not widespread—limited primarily to the United States. The Europeans, and in turn the former colonized countries of the developing world, took a significantly different route. They took the government owned and operated route. Only now is the U.S model beginning to spread into a few developing countries. One active player in this arena has been the U.S.-based National Telephone Cooperative Associations with modest support from the U.S. Agency for International Development (USAID).[4] But even here the efforts are minuscule compared to the overwhelming need.

And what is this model? A quick historical survey tells the story. When the telephone was first invented in 1876, it was patented in the U.S. As a result, the initial years were constrained to deployment by a single monopoly provider, Bell. Not government-owned, but government authorized. The results were virtually indistinguishable. Pricing was high, access very limited, growth was slow, and the monopolist extracted extraordinary profits (with Bell having an annual ROI on the order of 46 percent). Teledensity was low and limited only to higher-density urban areas. Virtually no service existed in rural areas. Many of these same conditions exist in developing countries today.

It was not until the basic patents on telephony expired in 1894 that telephony expansion took off like a rocket. Within but a few years growth rates of new lines grew from an average growth rate of 6 percent to an annual growth rate of around 26 percent. At the time Bell's monopoly ended in 1894 (e.g., when the patents expired), there were an estimated 270,000 lines within the Bell system, and these were virtually all concentrated in cities with little or no penetration in rural areas. Teledensity in the U.S. in 1885 was estimated at 0.2/100. It was 0.43/100 ten years later in 1895. By contrast, during the 13 years following the end of Bell patent, over 6,000,000 lines were in service. These were evenly split between Bell and the independent telcos. And no longer was service limited to high-density urban areas. Rather, it was reasonably-well distributed across the country, both urban and rural.

---

[4] See Marlee Norton. A New Day Dawns in Poland. Rural Telecommunications. May-June 1998. pp. 14

The number of telephones had grown by approximately 2,100 percent between 1894 and 1907.[5] Teledensity in the U.S. was estimated as growing from 0.43/100 in 1895 to 6.67/100 in 1906.[6] The rural state of Iowa had a higher teledensity than New York City—a location where Bell retained a virtual monopoly.[7]

While the free-market advocates put forward the notion that this represents the power of competition, in fact this is but one piece of the equation—and potentially quite misleading at that. In truth, head-to-head competition took place in very few locations throughout the U.S. The more plausible explanation for this explosion is the following: 1) open access to the technology—it was no longer restricted by the patent that had created a monopoly situation, 2) free market entry—nothing was in place to restrict telephone companies from starting a business—the FCC didn't exist for several decades later, and 3) hundreds of small telcos, including many local cooperatives, entered the market—actually thousands. By 1902, just eight years after the patents expired, there were 3,000 non-Bell commercial systems and three major manufacturers of telephone equipment supporting this rapidly growing industry.[8] The number of independent rural systems operating peaked in the U.S. in 1927 at over 6,000. On average, these telcos served small populations of users, typically with an average customer base of less than 200. Virtually none were interconnected.[9] This came later as technology advanced and when in 1913 Bell agreed (via the Kingsbury Commitment) to interconnect these independent companies, as well as those that they owned.

The Model

For developing countries today, the applicability and lessons-learned should be obvious. In summary, the most rapid expansion of telephony in the history of the world was largely a *massively parallel, bottoms-up, community-based strategy*. It took place at a time where the critical role of the government was simply staying out of the way; letting it happen. The dominant focus of this dynamic was very simple: establishing service for the First Mile First. While perhaps naive to think that any country can replicate exactly what took place in the U.S. a hundred years ago, the model is just as valid today as it was then— recognizing of course that there is the need for localized adaptation to achieve optimal results. In fact, there is every reason to believe that with today's more advanced technologies the success can be achieved

---

[6] Brock, Gerald W. The Telecommunication Industry: The Dynamics of Market Structure. Cambridge, MA: Harvard University Press, 1983, pp. 89-176

[7] Gabel, David and Weinman, David. Historical Perspectives on Competition between Local Operating Companies: The United States, 1894-1914. Queens College, CUNY-NYC,NY. 46pp.

[7] Mueller, Milton L. Jr. (1997) Universal Service: Competition, Interconnection, and Monopoly in the Making of the American Telephone System. MIT Press and the AEI Press

[8] Foundation for Rural Service. Independent Telephone Companies: Keeping Rural America Connected. 1996. Also see the FRS Internet Web Site at http://www.frs.org/indy.htm

[9] National Cooperative Bank (NCB). Every day in America 1.2 million people communicate on phone lines serviced by telecommunication cooperatives. http://www.ncb.com/day/article10.htm

substantially faster and at significantly lower relative costs. The potential? We haven't seen anything yet...if we would simply adjust the approach!

Today's Technologies

The most important technological advances in telecommunications that have direct application to this First Mile First model, in a word is "wireless." No longer do farmers have to string wire along rows of trees, fence posts, or install poles. Now an entire local system can be put into place in a matter of weeks without wires. Where needed, these systems can be powered by solar panels due to the low power requirements of micro-technology and digital signaling. A growing number of smaller stand-along systems are becoming available at a lower and lower cost per connection. These systems not only provide local switching, billing, operations, etc., but they also provide industry standard interfaces to current public switched telephone networks (PSTNs).

Wide arrays of technologies are being packaged into small-distributed switching systems that can be deployed to meet these local needs. These increasingly rely on the newer digital technologies such as Code Division Multiple Access (CDMA) and Digital Enhanced Cordless Telephony (DECT). The CDMA solutions service larger "macro-cells" with a single transceiver supporting up to 20-30 kilometer radius service area under ideal conditions. The DECT systems service smaller "micro- cells" of approximately 5 kilometers. But even the older analog systems and various Time Division Multiple Access (TDMA) based solutions (including GSM) are more than adequate, broadly deployed and with years of proven use behind them. One or the more promising near-future solutions is a phone-to-phone system being developed by Frank Neukomm at Direct Wireless of Houston, Texas.[12] It may be but a year away from production and promises to substantially lower costs and simplify deployment and operations even further.

For interconnecting these small rural systems, again wireless to the rescue. This can be in the form of higher-end microwave solutions for high-traffic areas, or stringing a series of lower-cost, solar-powered, repeaters together. More recently there has been a refocusing on satellites for this interconnection. Fabrice Langreney of Intelsat has undertaken promising pilots in Peru and more recently in Senegal---relying on AMPS and DECT for the local distribution and thin-route DAMA services off Intelsat satellites for linking these systems into the nations' PSTN.

Those that labored to put into place a small rural system a hundred years ago would not believe how easily these systems can now be put into place, and how trouble-free their operations have become.

> ***...experiences and observations from the author***
>
> While the technologies put forward by Frank Neukomm in the mid-late 1990s did not materialize, others have, with several holding promise for supporting small rural Internet deployments. Distribution solutions include Wi-Fi, Small Cells (4G/LTE and soon 5G), Open BTS, Open Cellular, and others. For the middle and first miles, such technologies as small microwave, point-to-point Wi-Fi, WiMAX, and TV White Space solutions also provide support. New generations of smaller micro-satellites, high altitude balloons, drones, etc., are just around the corner.

What's Holding it Back?

With known demand, a tried and proven model, and advanced technologies, the obvious question is, "Why isn't it happening?" Why are countries not taking advantage of these advances and moving forward? Is it insufficient demand? Not likely, with 70 percent of the nations' populations living in rural areas and frequently on the order of 50 percent plus/minus of the nations' GDP attributed to activities carried out in the rural areas. Is it the cost? No, in fact systems can be put in place now that bring the cost for a local system to under $1,000 per line. And this cost is dropping fast, likely to be in the $400-$500 range within a few years. While depending on average GDP in a given rural area, the economics are such that substantially more teledensity can be realized than what is now taking place. Is it lack of human capacity? Not likely as U.S. farmers were never considered the fountain of high-tech but still pulled it off—no reason to suspect it's substantially different in developing countries today. There is always some mix of creative entrepreneurs in every location, worldwide. Is it for lack of proven benefits? No, an increasing amount of data and experience shows the opposite to be the case. It is becoming understood that telecommunications is an enabling technology that contributes significantly to social and economic improvements in rural areas, perhaps even greater than roads and electricity. It was the case a hundred years ago; it's been the experience today.

Most likely it's not happening for any number of reasons, with history, tradition, and uncertainty high on the list. As governments seek to liberalize their telecommunications environment there is an understandable level of conservatism involved, with most reforms "inching" forward rather than "bounding" forward. When multiple players are allowed entrance into the local marketplace, as a rule it is in the cellular markets of urban areas as complementary services to existing basic landline services. When multiple carriers are allowed to operate it's once again either in the urban areas, but most typically undertaken to divide the country geographically amongst the players. And the number of players can generally be counted on one hand with several fingers left over. While this represents progress, it is most typically urban focused, involves few players, with some rural-oriented requirements included in the package. While this represents progress, it's unusually slow—with the *fast world* nations gaining more distance and advantage in the interim.

Making It Happen

The current dynamics taking place in many developing countries is clearly headed in the right direction—only typically not fast enough and not of significance such that near-term changes will take place in the rural areas. Clearly, adjustments are needed. Not as an alternative to the current liberalization, competition, transparency, enabling policy and regulatory reform, but rather as a complement to these efforts. The following adjustments can enhance the process.

Rethinking the Model—Today's model for reaching the population living in rural areas of developing countries is a built-out strategy; urban first, rural later. Even those countries taking aggressive actions to reach rural populations are simply placing one, two, perhaps a few more, telephones into a community. The orientation is based on a notion that it's more important for people living in rural communities to be connected to those in the urban centers than it is to be connected to their neighbors or a nearby village. Long distance service without local calling. It simply makes no sense. While interconnection is needed, a First Mile First approach places high value on servicing the local needs first, then connecting to the "outside" when the community determines this to be of high value. It empowers the local community to meet their needs and at the same time supports the national strategy for expanding coverage as rapidly as possible.

Targeted Policy and Regulatory Reforms—The need for telecommunications policy and regulatory reform is well understood by nearly everyone. Many countries, both developed and developing, are already headed down this path with multiple approaches being taken. It is also understood this is not quick nor without a lot of details, complexities, and political compromise. But the biggest issue, besides the chosen approach, is speed and timing of the ultimate sought-after goal. What is needed is a parallel path that comes alongside the more fundamental changes many countries need, one that allows near-term improvements in key areas such as rural populations. It is recommended that rural areas receive special consideration and be allowed to progress on a more aggressive parallel track by lowering entry restrictions for providers and granting licensing on a fast track.

Making Technology Available—During the late 1800s it was the end of the patent that triggered the explosion of telecommunications in the U.S.—especially in the rural areas. For many countries today, the current limitations on technology are the restrictive tariffs levied on importing high-tech. These price the technology such that it becomes unavailable. These barriers must be eliminated. The artificial constraint only slows progress even further. While building local in-country industrial capacity is important, and protecting emerging local businesses is also critical, to handicap the entire nation's ability to join the *fast world* is simply too high a price to pay.

License/Franchise Rural Systems—The second key lesson from the U.S. experience is the need for enabling regulatory conditions to allow for deployment and expansion of newly-available technology. In the 1880s, nothing stood in the way of deploying the telephony technologies. There were no restraints; there was no FCC. Today things aren't so simple, even in the U.S. While a growing number of countries are beginning to license/franchise wireless mobile systems, typically these serve only high-density urban areas. The same process needs to be applied to rural areas. Governments can provide incentives if needed, by establishing universal service/access funds, though this should be carefully undertaken. In those countries that are not on a fast track to privatize their monopoly carrier, operators should develop a franchising approach whereby independent rural operators are put in place to provide services in selected areas of the country. This approach off-loads the need for their direct involvement, yet extends the reach of the telecom company into un-served areas, thereby enhancing their value.

Manage the Build In—While a complete hands-off strategy could potentially work, more likely there is the need for oversight and order to expand telecommunications into rural areas. However, this should be dominated by empowerment-oriented policies and regulations with minimal artificial constraints. There is the need for managing radio spectrum. And while competition has its value, it is not without its downside and needs some oversight to ensure there is efficient market entry. Interconnection is a mandate for all players in the sector and must be supported not only with policy and regulation, but also some base level standards-setting to ensure the multiple products provide requisite integration. Revenue sharing is another area where government may need to establish guidance—the details to be worked out by negotiations between the parties themselves. If a licensing arrangement were deployed, then parsing out regions, areas of coverage, drafting model-licensing agreements, etc. would enable faster deployment. Tariff setting, international access, and dispute resolution, also need attention, but expansion can proceed without waiting for all the details to be worked out on broader telecommunications sector reforms.

...experiences and observations from the author

In addition to the emerging technologies suitable for small-scale rural deployments, there are currently innovative business models emerging that hold significant potential for supporting massively parallel roll out. These include Mobile Virtual Network Operators (MVNOs), Mobil Virtual Network Enablers (MVNEs), along with more established approaches such as Build Own Operate (BOO) and Build Own Transfer (BOT), where the small rural telecom or ISP can have direct support from and link to the larger MNOs.

There are also newer business models built around emerging technologies such as low-cost Wi-Fi, TV White Space, shared and unlicensed frequencies, etc. The Wireless Internet Service Providers (WISPs) are beginning to experience successes across the globe.

Establish Partnerships—With the richness of experience and the need to pull together resources from a wide array of contributors, it is essential to connect players into cohesive partnerships. This may include bringing in the equipment manufacturers to assist in design, implementation, and initial operations; it may lean on National Telecommunications Cooperative Association (NTCA) members to secure management experience to avoid common pitfalls. Clearly the government has a critical enabling role; bilateral and multi-lateral development institutions may become engaged in piloting or even helping to support a national approach; and assuming a supportive business climate, foreign investors can be brought into the mix. Going it alone is simply not effective, efficient, nor fast.

Leverage the Resource—Most rural areas in developing countries lack more than telephones. A telecommunications infrastructure provides a valuable resource for delivering other services frequently missing in these environments. While this approach will significantly increase teledensity, in most locations there is the need to consolidate services into community information/resource centers. Health clinics can use the infrastructure to expand services via tele-medicine. Businesses can leverage the technologies by improved linkages with customers and partners—expanding into e-commerce by marketing local products on the Internet. The centers can support educational activities via distance-learning and computer-based training. Governments can leverage the network to expand the reach of key government services to rural areas—what is now being referred to as "e-government." Agricultural support can extend to include current pricing and information on new research. The potential uses are unlimited, but it starts with connecting the first mile—making access available at affordable prices.

Summary

There is an opportunity to learn from the experiences of the past 100 years, and leverage these experiences with the more advanced technologies. But in doing so, it is essential that the model be carefully chosen and locally adapted. While the build-out model has been used in most parts of the world, a massively parallel build-in strategy with hundreds, even thousands of small independent telcos may be what is actually needed and more appropriate. Not as a mutually-exclusive "either-or" alternative, but rather as a complementary approach. It is also important that some "out of the box" thinking go into how best to apply newer technologies. While wireless was initially developed to satisfy the mobile demands of urban customers in developed countries, in fact it may have its greatest potential use in providing fixed access to low-density rural areas in developing countries where there are no existing wire line services. And while the ultimate goal is to create an integrated national network, the fastest approach may be through first interconnecting small independent systems to meet local demands.

As the developed countries look toward broadband access, with even faster adoption of high-tech solutions, the gap between the majority of populations living in the developing countries relative to access afforded by enhanced technologies is widening, not getting narrower. It is critical that new and creative approaches be adopted to close this gap. One of the most promising approaches to address this issue may not be so new after all.

...experiences and observations from the author

The above research paper, "Wireless Village: First-Mile-First", was written nearly 20 years ago. It reflected on the U.S. dynamic associated with the rapid rollout of rural telephony that took place more than 100 years ago.

The situation 100 years ago is oddly similar to the needed rollout of affordable Internet just starting to take place in the LDCs and LCCs. Unfortunately, this is most often undertaken at a small scale and in isolated localities. It should not be a surprise that where this is happening, it is typically happening with reliance on unlicensed Wi-Fi and at times, TV White Space technologies. This approach is in keeping with a key theme presented in this 1998 research paper—that of the success in part being the result of having minimum government oversight, where governments are not in the way of making progress, and where there is the least amount of regulation governing the deployments of Internet-related technologies.

In an odd twist while writing this book, even here in the U.S., the government and firms like Microsoft just recently launched new initiatives to expand affordable Internet in the rural U.S. As it turns out, the lack of rural connectivity is not limited to the LDCs and LCCs.

CHAPTER 2

Dynamics of 2000-2015

The 2000-2015 timeframe was rich in heightened global focus on expanding affordable access to voice and Internet services. This timeframe corresponds with the 15 years covered by the Millennium Development Goals (MDGs). Within this construct, there was an increased emphasis led by the UN's International Telecommunications Union (ITU).

The ITU provided needed focus, advocacy, and tracking of progress on a wide array of ICT-related topics and measurements. The countries complemented this themselves, where there was a major thrust in liberalizing their telecommunications environment and opening it to competition, especially in the mobile sector.

With the liberalized market, the commercial mobile carriers rapidly expanded their networks. The multilateral and bilateral development community also heightened their engagement in this space. And towards the end of the MDG timeframe, the international private sector high-tech firms became actively involved in expanding ICTs through public-private partnerships.

Millennium Development Goals (MDGs) – 2010 - 2015

The MDGs[10] reflected eight international development goals that came out of the Millennium Summit in 2000, as adopted through the Millennium Declaration. All 189-country members of the United Nations (UN) adopted these goals and targeted 2015 as their planning horizon. Collectively, the goals for 2015 helped focus international development, not only of the UN organizations, but other multilateral and bilateral development institutions.

> **Millennium Development Goals – 2015**
> 1. To eradicate extreme poverty and hunger
> 2. To achieve universal primary education
> 3. To promote gender equality and empower women
> 4. To reduce child mortality
> 5. To improve maternal health
> 6. To combat HIV/AIDS, malaria and other diseases
> 7. To ensure environmental sustainability
> 8. To develop a global partnership for development

Within the MDGs, leveraging Information and Communications Technologies (ICTs) to support the international development agenda had virtually zero mention. However, the UN's International Telecommunications Union (ITU) augmented the MDGs by championing a major international thrust on ICTs.

---

[10] http://www.un.org/millenniumgoals/

World Summit on Information Society (WSIS) Targets for 2015

The ITU's strategic focus on ICTs took shape through two WSIS meetings, one held in 2003 in Geneva, and the other held in 2005 in Tunis. The second Summit concluded with the establishment of specific targets for 2015 that corresponded to the MDG timeframe. While the WSIS Plan of Action did not initially attach quantitative indicators to these targets, a measurement framework was subsequently established with periodic statistical reporting as to the progress being made.

There were initially 10 WSIS Targets establish in 2005, with an 11th added later.

WSIS Targets – 2015

1. To connect all villages with ICTs and establish community access points
2. To connect all secondary schools and primary schools with ICTs
3. To connect all scientific and research centers with ICTs
4. To connect all public libraries, museums, post offices and archives with ICTs
5. To connect all health centers and hospitals with ICTs
6. To connect all central government departments and establish websites
7. To adapt all primary and secondary school curricula to meet the challenges of the Information Society, taking into account national circumstances
8. To ensure that all the world's population have access to television and radio services
9. To encourage the development of content and to put in place technical conditions in order to facilitate the presence and use of all world languages on the Internet
10. To ensure that more than half the world's inhabitants have access to ICTs within their reach and make use of them
11. Connect all businesses with ICT

Broadband Commission for Digital Development

In 2010 the ITU and UNESCO set up the Broadband Commission for Digital Development to advance the UN's efforts to meet the MDGs through an expanded focus on digital development. The Secretary General's call for action was in direct recognition that ICTs role was critically important in supporting the achievement of all eight of the MDGs.

The Commission initially established four Broadband Targets for 2015, and then in 2013, established a 5th target addressing gender, with a 2020 target date.

Broadband Commission Targets – 2015

1. Making broadband policy universal—by 2015, all countries should have a national broadband plan or strategy or include broadband in their Universal Access/Service Definitions
2. Making broadband affordable—2015, entry-level broadband services should be made affordable in developing countries through adequate regulation and market forces (amounting to less than 5% of average monthly income.
3. Connecting homes to broadband—by 2015, 40% of households in developing countries should have Internet access
4. Getting people online—by 2015, Internet penetration should reach 60% worldwide, 50% in developing countries and 15% in LDCs

In 2013 the Broadband Commission established a 5th Target—this one addressing gender equality with a 2020 target date.

5. Achieving gender equality in access to broadband by 2020

The Broadband Commission has actively collected and published annual reports on the progress of their now-five targets.

...experiences and observations of the author

It is important to note that in the WSIS Targets, the dominant emphasis was aimed at "Connect," while the more recent Broadband Commission's focus was on "Broadband." That said, with both the WSIS and BBC Targets, there are specific references to "services" to be connected— community centers, schools, research centers, hospitals, governments, and households.

The additional attention on connectivity in the mid-2000s reflects the on-the-ground reality at the time, whereas a few years later, with the Internet, expanding affordable access was the important component. But it's also worth noting that even back in the mid-2000s there was a focus beginning to emerge on priorities from where the value was being targeted.

Ratchet forward to today, 2017. There has been considerable progress made by the developing countries—though the LDCs continue to lag in the deployment, affordability, and adoption of the Internet.

Commission on Science and Technology for Development (CSTD)

The CSTD is a subsidiary organization of the UN's Economic and Social Council (ECOSOC), with UNCTAD serving as its Secretariat. Established in 1993, the dominant focus of CSTD includes: 1) the examination of priority science, technology, engineering and innovation (STI) issues, including ICTs, and their implications for development; 2) the advancement of understanding on science and technology policies, as well as sharing of best practices, particularly for the benefit of developing countries, and; 3) the formulation of resolutions for implementation by ECOSOC, including recommendations and guidelines for S&T and engineering, ICTs for development, and related innovation within the United Nation's system. In 2017 there were 43-member States, each serving 4-year terms. The U.S. has been a permanent member since 2007.

The CSTD has two critical responsibilities relating to the above discussion, including: 1) its ongoing advocacy for leveraging STI and ICTs for the advancement of the UN's MDGs; and 2) assisting ECOSOC as the focal point for assessment of progress and follow-up to outcomes of the WSIS Action Lines.

The Commission has devoted an increasing amount of attention to assessing the progress of the 2015 MDGs and WSIS Action Lines. In addition, the CSTD has explored the evolving Sustainable Development Goals and Post-2015 development agenda and the seminal role of STI and ICTs.

...experiences and observations from the author

Between 2012-2015, I had the opportunity to work with the U.S. State Department and participate in the annual meetings of the United Nation's CSTD in Geneva, as a panel moderator or as a speaker. I also participated in dialog with State and others to get the then-emerging SDGs to focus some level of attention on ICT4D topics. And while to a degree this was achieved through SDG Target 9c, most of us working in this space were disappointed with the minimal mention of ICT at the SDG level.

We were however, able to influence CSTD to conduct further research that resulted in two reports—Digital Development and Foresight for Digital Development.

Reference:
http://unctad.org/meetings/en/SessionalDocuments/ecn162016d3_en.pdf

...experiences and observations from the author

While this book focuses on Digitization and the challenges associated with the Least Developed Countries, my experience has also revealed that rural broadband challenges are not limited to the LDCs.

Shortly after forming the GBI Program at USAID, as we were preparing for launch, I received a call from the U.S. Federal Communications Commission (FCC). They had heard about our efforts and wanted to know more. We met with them, and the theme of their interest was built around the need to do something similar in the United States. Not long after this dialog, the White House and the Department of Commerce announced and launched a rural broadband program in the U.S.

Another example of growing interest in rural broadband occurred in May 2012 at the annual CSTD meeting. I moderated a panel on the topic of expanding rural broadband. Afterwards I had a short conversation with the then-U.S. Ambassador to the UN in Geneva. She indicated her pleasure of our panel discussion. She went on to mention that immediately after our session, two country representatives approached her and requested all our materials from the five panelists (from the U.S., Pakistan, Colombia and Ghana). The two CSTD members who approached her were from China and Russia.

While the LDCs acutely need affordable Internet in their rural communities, expanding rural Internet is an issue that many of the largest countries, even the developed countries, are also trying to address, including the United States, China, and Russia. Another example is India. Although India is a Least Connected Country (LCC), not an LDC, India has made rural Internet connectivity a priority. This was most recently captured as part of their "Digital India". In late 2015, I received an invitation to assemble a panel and speak at a connectivity-related event in Delhi supporting their "Digital India" agenda.

USAID-Related Engagements

During the timeframe of the MDGs, many bilateral development organizations and agencies engaged in supporting the ICT4Development (ICT4D) space. In fact, a few of these were underway well before 2000 when the MDGs were established.

In that the bulk of my engagements were associated with the U.S. Agency for International Development (USAID), my focus, and the content that follows, centers around USAID. However, other countries, through their international development agencies, were also engaged. These include Britain's DFID-UKaid, Sweden's Sita, Australia's DFAT, etc.

For USAID, the involvement in expanding connectivity is what I would characterize as a sequence of "episodic," initiatives—each established based on a specific focus at the time the initiatives were developed. These typically lasted on the order of 5-6 years, with multi-year gaps between the end of one initiative, and the launch of the next initiative. The following reflects several USAID's ICT-related initiatives, including several where I was engaged at some level. This "initiative" orientation contrasts with what can be thought of as a more permanent "program" construct—where there is an ongoing focus (e.g. education, health, economic development, agriculture, environment, etc.)

USAID's Leland Initiative (LI)

In the mid-1990s, USAID's Africa Bureau launched an initiative that focused on introducing the Internet into approximately 20 countries in Africa. This five-year plus initiative was launched in June 1996, and named after Representative Mickey Leland, a strong advocate for U.S. engagement in Africa. The focus was threefold: 1) create an enabling environment, 2) encourage the establishment of a sustainable support of Internet access, and 3) use the Internet as a tool for sustainable access.[11]

USAID's Programs Leveraging ICTs

USAID's programs have a history of incorporating ICTs into their projects—be it education, health, agriculture, etc. In 1998, USAID's then-Bureau of Policy and Program Coordination (PPC) undertook an internal study to determine the amount of ICT-related work imbedded across USAID's programs. This analysis showed that there was on the order of US$ 300M ICT-related investments made annually across USAID programs. At that time the overall Agency budget was on the

---

[11] http://www.itu.int/ITU-D/ict-stories/themes/case_studies/leland

order of $7 Billion. As a percent of USAID's total program budget levels, this amount was relatively consistent with the ICT investments in private sector corporations.

Today, that amount would be considerably higher since USAID's program budget has grown substantially in the 2000s (on the order of 4X) from when this study was done. In addition, in recent years ICTs are increasingly viewed as having a higher priority.

USAID's Program Technology Transfer (PTT)

In the late 1990s, USAID's Information and Resource Management (IRM) Office implemented the PTT. The focus of the PTT was to provide technical support and expertise to the Agency's Missions with regards to imbedding ICTs into their programs and projects.

The PTT's mandate was to review Agency projects where there was an imbedded ICT component that totaled over $100K. It was not uncommon for the PTT team to review on the order of 300+ projects every year.

Another form of PTT support provided more in-depth support to the Missions via in-country consultations and assessments. Where the assessments identified potential for ICT-related activities that the Missions wanted to pursue, the assessment team would later return and work with Mission staff to develop a detailed project design, schedule and budget.

Typically, the Missions provided the project funding, with the PTT Team providing the needed expertise, project management, and execution through existing central agency contracts. In other situations, the PPT developed specifications with the implementation imbedded into an existing or planned Mission contract.

...experiences and observations from the author

When I retired out of USAID in the fall of 1999, I consulted on a number of USAID projects through several contractor firms. This included being part of a team undertaking an Internet for Economic Development (IED) assessment in Morocco. Another Assessment in Sri Lanka followed this, and in early 2000, there was an assessment in the Democratic Republic of Congo (DRC). In mid-2000 I spent nearly 4 months in Egypt designing a nearly $40M ICT-related program for USAID's Mission in Cairo.

The assessment process consisted of a structured 1-week desk study to gather core data and contacts prior to getting on the plane. From there our team went in country for two weeks+ and had a report pretty much drafted by the time we returned to Washington DC. The report was a structured analytical document that concluded with a preliminary set of opportunities that supported a Mission's country development portfolio. Often, we'd include a preliminary design. This structured approach also incorporated an out-debrief to the USAID Mission management and staff.

The structure of the assessment and report followed what we referred to as the 5Ps – Pipes (telecoms infrastructure); Public Sector (policy, regulatory, and use of ICTs by the government); Private Sector (the local business environment, status of the ICT sector); People (looking at the ICT skills level, education, ICT in education, higher education ICT programs, etc.); and Program (what did USAID's country-level development portfolio look like and where was there the potential opportunities for better leveraging ICTs). We also included in our assessments what other donor programs were doing in the country to explore opportunities for synergy between the donors in the ICT space.

During the early-to-mid 2000s, I was heavily involved in conducting country assessments. This included working as a team member, or serving as team lead for several assessments in many countries—Armenia, Romania, Georgia, Eritrea, Sri Lanka, Vietnam, Cambodia, Laos, and Indonesia. I became most engaged over an extended period in Armenia and Eritrea because of the assessments done in those two countries.

In reflecting on these experiences, I find it interesting that while expanding connectivity was a major theme of the WSIS in the early 2000s, our analytical approach is what could be viewed as a precursor to what is now a much broader, refined, and quantified state-of-the-art, "Digitization," the model reflected in the second bookend of this book.

...experiences and observations from the author

One of the PTT country programs where I had the longest and most intense involvement was in **Armenia**. The initial assessment took place in mid-2000, where we found a rich opportunity for multiple ICT-related engagements to add to the Mission's development portfolio. The Mission provided funding on the order of $500K-$2.4M for each of the next 5+ years—through 2006. Major initiatives included:

National ICT Strategy—shortly after the Mission buy-in, we worked closely with the Armenian government and the World Bank in developing a national ICT strategy. This included setting up a support structure at the highest levels of government, as well as the private sector, with buy-in from both. After 18 months, progress was assessed that led to an updated plan as an outcome.

Digitizing Financial Capabilities—the default situation in 2000 was that the country was a pure cash-based society. Over the years, we worked with the Central Bank of Armenia (CBA) and the local private banks and moved it to where it included international electronic financial transactions, automated auditing of the CBA, electronic daily reconciliation with the banks, introduction of a national debit card system, ATMs, and introduction of locally-issued Visa and MasterCard credit cards.

Expanding ICT Education-Skill Building—we established Cisco Academies in three State Universities, along with computer labs. This included supporting students getting their Cisco certifications. Later we established Master of Science in Information Systems (MSIS) programs in these three State Universities, each specializing in a specific area of study.

Supporting Competition in the Telecom Sector—the default telecom position was an ArmenTel monopoly, with the company, OTE, owning 85% and the Armenian government 15%. We undertook a study surfacing the fact that ArmenTel was not living up to its license obligations for expanding rural coverage. The Minister of Justice took ArmenTel to court. This led to opening up the local mobile market. We undertook the development of new legislation that supported competition and developed specifications for capacity building within a new joint utility regulator.

Supporting the Growth of their High-Tech Sector—another key focus was supporting the local high-tech sector. Under Soviet era, high-tech support had been significant in Armenia, but it was nearly zero in 1990. We brought in international trainers and developed strong linkages between the Armenia sector and a network of high-tech Armenian Diaspora in the U.S. This relationship was instrumental in expanding knowledge transfer to the local firms and entrepreneurs, as well as opening business opportunities for several U.S. firms to establish local operations and businesses.

USAID's Last Mile Initiative (LMI)

The LMI[12] was established in 2004 as an augmentation to the Program Technology Transfer (PTT) program. The dominant focus of the LMI was to explore opportunities for closing the Urban-Rural digital divide. The LMI focused on four key areas:

1. Policy and legal/regulatory environment,

2. Advanced information technologies,

3. Application within USAID's Missions' development portfolio, and

4. Public and private sector partnering.

The LMI had varying levels of engagement in over 30 countries before it wound down in 2008.

...experiences and observations from the author

My engagement with the LMI started with a pre-assessment in the South American country of Peru. This was just prior to the formal launch of the LMI. The trip to Peru explored and field-tested the assessment approach.

Once the LMI was officially launched, I became involved in leading LMI assessments and preliminary designs, and subsequently engaged in the final project design for several countries. In addition, I managed or supported many country-level projects. These included: Vietnam, Mongolia, Cambodia, Laos, Sri Lanka, Georgia, and Romania, among others. I subsequently managed LMI engagements in Vietnam and Mongolia, and worked on other country-level LMI projects as part of a larger team.

In an interesting twist, after the LMI project ended, I learned from a former USAID colleague that my 1999 Wireless Village: First-Mile-First research paper served as the foundation upon which he designed and obtained funding for this Agency-wide LMI connectivity initiative. A small world.

While the LMI had at least preliminary exploration and assessment engagements in nearly 30 countries, key successes were achieved in Macedonia, Vietnam, Mongolia, and Sri Lanka.

---

[12] http://mikeb.inta.gatech.edu/LMI_files/LMI.ebook.pdf

...experiences and observations from the author

The following highlights the major LMI successes.

Vietnam LMI—my engagement in the LMI in Vietnam was from start to finish. The project began with an LMI Assessment where we settled on two areas of engagement: 1) getting a newly-approved Universal Service Fund (USF) up and running, and 2) demonstrating that there were emerging technologies that could provide wireless broadband into rural communities. We got the USF up and running within about a year, with an initial annual disbursement of US$ 54M. This took place in the mid-2000s. Recent 2016 data reflects that the USF has collected and disbursed well over US$ 1B since the initial launch. A key innovation we introduced in Vietnam was that a portion of these funds would be set aside as concessionary loans to support build-out—not as a subsidy, but rather as low-interest investment loans to be paid back, and made available to loan out again. Our rural deployment work was in partnership with Intel who deployed two WiMAX networks.

Mongolia LMI—my engagement in Mongolia included an initial assessment and design. The preliminary design approach was similar to the one in Vietnam, only the focus was on rural telephony and setting up their USF. As we moved towards implementation we collaborated closely with the World Bank—who took on designing and launching the USF. We concentrated on supporting an initial deployment of four low-cost rural soum-level networks. A soum is a small rural village. Our support included technical assistance, a small amount of capital for equipment, and a business-financial model. All four of these deployments were financially sustainable within 3-4 months. The local firm we worked with, Incomnet, now has over 200 rural deployments.

Sri Lanka LMI—in Sri Lanka the LMI focused on building a network of small tele-centers called EasySeva. This project pioneered new financing and business models in partnership with local government and private sector (both local and international). A network of 55 small centers was established, with ongoing support provided via Dialog, a local MNO.

Macedonia LMI—The MK Connect initiative was built around USAID's education program and included placing computer labs, complete with curriculum, teacher training, and Internet access, in over 450 schools, country wide. The project was innovative and committed to purchasing Internet access for two years from any firm that could provide national coverage. A local ISP obtained financing based on USAID's guaranteed future income stream, and built a nation-wide wireless broadband network. It was the first nation ever to have nation-wide wireless broadband coverage. The wireless technology was Motorola's Canopy product set, a pre-WiMAX technology.

...experiences and observations from the author

The two countries where I had the most direct engagement within the LMI program were Vietnam and Mongolia. This occurred in the mid-2000s. We designed into each project a small number of rural deployments, both relying on VoIP voice solutions. In Vietnam, we also provided full Internet access via WiMAX. Both relied on Satellite for some of the rural deployments. The following two experiences occurred in the initial deployment phase and are included to drive home the reality that even 10 years back it was possible to expand sustainable services to rural populations.

Vietnam—There were two WiMAX deployments in the Lao Cai Province of Vietnam, one in the capital and the other in the small village of TaVan. Both were serviced by a VoIP server located in Hanoi that provided full access to their PSTN and MNOs, as well as to a gateway for access to the Internet. Intel was the major U.S. private sector partner with USAID. I recall one day when my wife and I were on a trip from Sunriver (Oregon) where we live, to Portland. About two-thirds of the way there, I received a call on my iPhone. It was Bernd, a colleague who worked for Intel, our key partner. The call was perfect, little delay. He gave me a local phone number so I could test calling him back directly. I did, and the call went through without a glitch. It was great to hear school children in the background and realize that this was, in all likelihood, the first ever call into and out of TaVan, Vietnam. TaVan is a village with a rich trekking tourism business, so those visiting the village, as well as the villagers immediately adopted the Internet access.

Mongolia—A similar situation occurred in Mongolia as the deployment was beginning. We eventually completed four deployments in rural soums in partnership with Khan Bank, and with Incomnet doing the deployments. Here too, the network was designed to integrate with the local PSTN, MNOs and International carriers. One Sunday evening I was in Sunriver when my iPhone rang. There was a strange string of digits on the screen, but I answered it anyway. David Elsmore, a contractor from the U.S. was on the other end of the call. He had been instructed that when the network was up and operational, the first phone call he was to make was to me. He called me from the small soum of Dadal. Not having any idea where Dadal was located, we ended the conversation with David cajoling me to look it up. I did. Dadal is claimed as one of the traditional birthplaces of Genghis Khan. Here again, after hanging up it hits you—we can connect the people of this planet on a sustainable basis. And when David's team later left the village, the elders lined up to give each team member a hug, thanking them for bringing the phone service to them so they could connect to their family and friends.

USAID's Global Broadband and Innovations (GBI) Program

The GBI[13] program launched in the fall of 2010. Its primary focus was to work with the country-level public sector in supporting the development of National Broadband Plans (NBPs) and the disbursement of Universal Service Funds (USFs). The GBI program also worked with local and international private sectors to catalyze the introduction of low-cost, off power grid, scalable, and replicable rural broadband deployments.

The initiative was designed and launched within USAID's Economic Growth, Agriculture, and Trade (EGAT) Bureau—later restructured and renamed the Economic Growth, Education, and Environment (E3) Bureau. The GBI program had successful engagements in Kenya, Nigeria, Ghana, Botswana, Colombia, Peru, Jamaica and Indonesia. It also teamed with Intel in supporting over a dozen regional and sub-regional workshops aimed at advancing the build-out of the Internet, with Education as a primary socioeconomic focus.

...experiences and observations from the author

Around 2009, USAID requested that I migrate away from my independent consulting activities with private sector firms and consult directly with the Agency. This request centered around a new role to help USAID design a connectivity initiative that built off the successes of the LMI. This change led to the launching of a new Global Broadband and Innovation (GBI) program beginning October 2010.

For most of 2011-2015, I provided direct consulting to the Agency in support of the GBI program—relocating back to WDC during 2011-2013 to help launch the GBI. The GBI program also served as the foundation for the design and launch of two other connectivity-oriented initiatives—the Broadband Partnership of the Americas (BPA), and the Africa Broadband Partnership (ABP). Both were derivatives of the GBI Program.

---

[13] http://unctad.org/meetings/en/Contribution/USAID%20Lab%20-%20GBI%20Program.pdf

...experiences and observations from the author

The GBI Program was designed, approved, and funded, based on a simple reality. There was growing awareness that in general, Universal Service Funds (USFs) often didn't work well. They tended to under-perform, if they performed at all. Two studies quantified this, one study undertaken by the ITU and one by GSMA. The studies revealed that the amount of undisbursed funds was on the order of US$ 15B. Due to our earlier success in Vietnam, we took on the challenge of tackling USFs. We met with the USAID Administrator and laid out a plan that if he would fund the GBI Program, we'd commit to releasing at least US$ 500M. When the program concluded, we had reshaped or released for disbursement, well over US$ 1.5B in USFs from several countries.

A critical component leading to the success of the GBI Program with regards to the USFs, was having a comprehensive and supported National Broadband Plan (NBP). While many countries have them, all too often the plans exist simply on paper. They are often developed without collaboration or support from the public and private sectors. Nor are the plans actively managed and pursued. It is essential for the NBPs to have broad support and be actively managed. These NBPs then become the blueprint for guiding USF disbursements.

For achieving success, these "Broadband" Plans are not just broadband plans—they should be thought of as "Digitization Plans," where all-things-digital are incorporated into the plans. One can think of it in this manner: the focus on Broadband was consistent with the MDGs (2000-2015), but we have now reached a level in most countries where the focus must expand beyond infrastructure to include leveraging the infrastructure—again, "Digitization." As reflected in the title of this book, this challenge still needs addressing, especially in the Least Developed Countries.

The GBI Program design also included engagements to support targeted deployments of low-cost, low-power wireless broadband networks. USAID partnered with Microsoft for the deployment of several TV White Space networks—one in Kenya (Mawingu), along with others in Botswana, Jamaica and Indonesia. These activities were not "pilots," but rather, our support focused on kickstarting initiatives that held promise of sustainability and scale of locally owned and managed networks.

In Kenya, the Mawingu project had support from a private company, local government, Microsoft, and funding from Jim Forster, Paul Allen's Vulcan Capital, and OPIC. In Jamaica financial support included funding from their USFs and FLOW, a local carrier. In Indonesia financial support included funding from their USO Fund and the carriers.

U.S. Global Development Lab

In the spring of 2014, less than 2 years before the MDGs came to an end, USAID formed the U.S. Global Development Lab with a focus on increasing the application of science, technology, innovation and partnerships to accelerate the Agency's development impact in helping to end extreme poverty and promote inclusive economic growth. The "Lab" was structured around several Centers, including; Development Research, Digital Development, Development Innovation, Transformational Partnerships, and Agency Integration. The Lab is supported through two Offices—Evaluation and Impact Assessment, and Engagement and Communications.[14]

In the fall of 2014, the Global Development and Innovation (GBI) program was relocated into the Center for Digital Development, as part of the Digital Inclusion (DI) Team. Over the following year the GBI program underwent a refresh to better align its business model with the focus of the Global Development Lab. In 2016 the Lab ended the GBI program and replaced it with Digital Inclusion's Connected Programs initiative.

Alliance for Affordable Internet (A4AI)

Shortly after the formation of USAID's Global Development Lab, USAID partnered with the World Wide Web Foundation in creating the A4AI.[15] This is a broad coalition of partners and contributors working to enable affordable access everywhere to advance life-changing power of the Internet. More than 80 diverse member organizations from around the world support this coalition, from civil society and the public and private sectors. USAID was one of the founding sponsors of A4AI.

A4AI's stated approach for achieving their target of reducing Internet access costs is:

> *"Though technological solutions to drive down costs and expand access are well advanced, their effectiveness to do so continues to be hamstrung by bad or outdated policies. Our work is centered on the belief that policy and regulatory reforms are the best tools to unlock technological advances, reduce the cost to connect, and enable universal access. A4AI builds local multi-stakeholder coalitions that identify priority local issues and work through a combination of advocacy, research and knowledge sharing to drive the policy changes necessary to reduce prices."*

---

[14] https://www.usaid.gov/GlobalDevLab/about
[15] https://en.m.wikipedia.org/wiki/Alliance_for-Affordable-Internet http://a4ai.org

Additionally,

"All of A4AI's members have agreed on a set of policy and regulatory good practices that guide our advocacy and on-the-ground work. These practices operate on the understanding that open, competitive and innovative broadband markets are key to reducing connectivity costs for operators and for consumers. A4AI recognizes that some populations will remain beyond the reach of the market, and so they also advocate for public access solutions (e.g., community Wi-Fi, free or subsidized access in public schools and community centers) to connect women, the poor, and other marginalized populations that might still be unable to afford an Internet connection."

To date, A4AI has engaged in Dominican Republic, Ghana, Liberia, Mozambique, Myanmar and Nigeria, with more country-level engagements being planned. In addition to country-level engagements, the A4AI has also generated a series of annual Affordability Reports, the latest report done in 2015/16.

...experiences and observations from the author

In addition to my engagements mentioned previously (the PTT, LMI and GBI), during the 2000-2015 timeframe I also had opportunity to work on other ICT-related projects that were components of larger country-level programs. Following are a few of those.

Egypt—in 2000 I spent three months in Cairo with a contractor team supporting USAID's economic development program, where ICT was thought to hold potential. This work focused on automating the Egyptian Ministry of Communications, capacity building of its telecom regulatory body, and private sector capacity building on a range of ICT-related topics. My efforts focused on assessing and identifying the priority needs, identifying local support, and designing a series of interventions. In 2000, the USAID program in Egypt was one of the largest programs funded and managed by USAID. The ICT-Program our team designed was on the order of US$ 40M to be carried out over several years.

DRC—in early 2000 I spent two weeks with a colleague in the Democratic Republic of Congo (DRC) with no significant results. In June 2010, I was asked to return to DRC with a specific focus on finding a solution to thwart the destruction of small remote villages by the Lords Resistance Army (LRA). In 2009 the LRA annihilated several villages and the world knew nothing for three weeks. My colleague (Troy Etulain) and I put forward the potential value of a solar powered satellite-based small-cell solution. USAID supported a trial in four remote villages. The early warning and communication system used mobile phones and low-cost cellular towers to connect the remote villages to the outside world so that villagers could immediately call for outside help whenever they noticed a possible threat. In addition to potentially saving future lives, a key outcome from this public-private partnership was that Vodacom, the implementing partner, learned from these initial four deployments, that they could make significant profits from servicing remote villages. By the end of 2015, Vodacom had extended services from the initial four villages to 800 small remote villages.

For more information see:

https://www.usaid.gov/results-data/success-stories/strengthening-community-protection-through-mobile-phone-coverage and

http://www.intelsat.com/wp-content/uploads/2016/03/Delivering-rural-cellular-services-in-DRC-Vodacom-7251-CS.pdf

...experiences and observations from the author

...and a few more

Jamaica—in the early 2000s, Jamaica was in the process of removing Cable and Wireless as the monopoly provider, drafting and gaining approval for new legislation to liberalize the local Telecom sector, and issuing tenders for two new mobile operators. I became engaged in a couple key areas, including: 1) working with the Ministry of Telecom, with support from the U.S. State Department, to review and comment on the draft legislation, and 2) working with the U.S. Federal Communications Commission (FCC). The FCC sent a team in country for two weeks with spectrum monitoring equipment. They assessed frequencies in use, to ensure that frequencies being licensed to the soon-to-be awarded mobile operators were open, and that no one was squatting on those frequencies. The result of this project opened the mobile market for competition. There was new Telecom legislation signed into law, and two mobile awards made, with licensing provisions requiring island-wide build-out within a set timeframe.

OECS Countries—for several years in the mid-2000s, I was under a consulting agreement to provide capacity building to the five English-speaking countries of the Dominica, Granada, St. Kitts and Nevis, St. Vincent and the Grenadines, and St. Lucia. The monopoly of Cable and Wireless was coming to an end in the Caribbean, and OECS and its member states had decided to set up a sub-regional regulatory authority in St. Lucia (ECTEL – Eastern Caribbean Telecommunications). ECTEL established a common direction for these five countries. Each country had their local Regulatory Authority. ECTEL consolidated sub-regional market power to attract new telecom firms to make local infrastructure investments in these countries. This is a possible model for smaller LDCs within a geographic region.

Tunisia—in the post Arab Spring era in Tunisia, I did a quick trip to Tunis with a couple colleagues to explore developing a program for aspirational, unemployed youth, many with college degrees. This project resulted in launching an effort that has proven successful in furthering Tunisian technical skills, mentoring and placement into jobs as the country serves as a regional hub.

Liberia—in 2013 the Ebola Crisis hit Western Africa, and USAID sprang into action to help address the issue. In the Telecom space, the GBI Program engaged with an on-the-ground Rapid Assessment. The assessment identified key areas of focus, and NetHope, one of the GBI Program implementers, pulled together several partner firms from the high-tech sector to address the immediate connectivity issues associated with rural health clinics. Other topics included supporting the Liberia Telecom Authority with spectrum management equipment and capacity building, and placing a fiber cable around the capital of Monrovia in partnership with Google and their partner, C-Squared.

Key Trends and Dynamics: 2000-2015

By virtually any measurement, these 15 years have been staggering with regards to the fundamental changes in "all-things-digital". At times we don't appreciate the progress that has been made. ITU data shows expansion across several measurements.

Increased Access

The following chart provides a quick capturing of several key dynamics that will serve as the base for moving forward in the 2015-2030 timeframe—the focus of the next chapter. The chart is from the ITU and displays *Individual Internet Access* data for years 2000-2015, with projections through 2021.

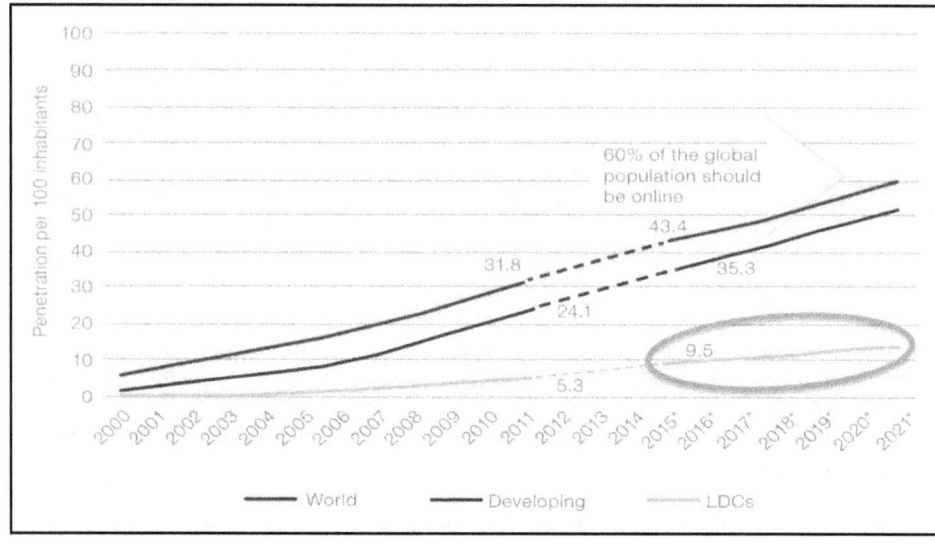

Individual Internet Access

The rapid growth in Internet access is the result of several factors:

- **Fruits of Deregulation**—The introduction of competition into the marketplace was largely responsible for the wireless explosion. Often this included a shift from a government owned-and-operated monopoly to privatization and introduction of competition into the marketplace.

- **Landlines to Wireless**—Another key contributor to this increased expansion was the introduction and upsurge of mobile technologies. Landline expansion has been virtually flat.

- **Mobile's Voice to Broadband**—Early in this timeframe broadband was pretty much limited to fixed line connections. The technology advances in the mobile space went from 2G-2G+ to 3G to 4G/LTE. And phones went from "dumb" phones to "smart" phones.

- **Fiber Build Out**—There was also an explosion of fiber build-out across the globe, the most notable being undersea fiber. In the late 1990s-early 2000s Africa was only marginally served—satellites being the dominant (and expensive) international linkages. Today there are often multiple fiber landings in a growing number of coastal countries. And there has been regional backbone build-out off these coastal landing points—serving both the coastal countries as well as neighboring landlocked countries. Currently the high capacity middle-mile and distribution into rural communities are often the key missing ingredients.

While progress has been significant, it is critical to note that the earlier graph reflects one exception where a low growth rate persists in accessing the Internet. The exception is the LDCs. The growth rate in the LDCs is slow and the gap between other developing countries and the LDCs continues to get wider.

Shifting Focus from Voice to Internet

At the beginning of this 2000-2015 period, voice was dominant, and the Internet was just starting to come online. Today the dominant focus is on affordable Internet, recognizing that the Internet Protocol (IP) has progressed to the point where most backhaul for voice is over IP—with CDMA and TDMA being the distribution protocols for mobile networks. And these mobile networks are increasingly delivering Internet via 3G and more recently, 4G/LTE. The emerging 5G will be online within the next few years.

- **Dumb Phones to Smart Phones**—With the expansion of the Internet and mobile access, the dumb phone capable of handling voice and texting (2G and 2.5G), has now given way to smart phones with a greater connectivity and processing capacity than desktops and laptops of the 1990s.

- **Desktops and Laptops to Smart Phones and Tablets**—And in a similar manner, desktops and laptops have given way to the smart phones and tablets with regards to the numbers of users.

- **Local Servers to Cloud Services**—With the expansion of the Internet over these 15 years, there was also a significant shift away from a reliance on in-house servers available through a local area network, to shared cloud services available off the Internet. This expands availability while at the same time lowers costs. It also opens up new business models across the globe.

International Development Community's Engagement

Over the 15 years of 2000-2015, the donor community expanded its involvement in ICT-related activities. This was reflected in a wide-range of engagements in both connectivity and application related themes. The bilateral development agencies actively supported the expansion of Internet access at some level. But even more than access, these organizations increasingly engaged in the value-added applications of mobile financial services, health-related services, education, addressing the refugee crisis, mapping, among others, within their more traditional development programs.

Cloud-Based Value-Added Services

With expanded availability of affordable Internet, focus is increasingly placed on value-added services—services that are available virtually anywhere via wireless broadband on a smart device. A whole new cloud-based industry has surfaced, especially in the last 5 years, which will drive this further as we go into the next 15 years.

Private Sector Cloud Players Engaging in Expanding Internet

The dynamic of cloud services has become dominant to the extent that the leading cloud service providers are increasingly becoming key players for expanding affordable Internet access. The logic is simple—to grow their respective businesses, they need to increase their customer base beyond the current more mature markets. Think Facebook, Microsoft, Google, and Amazon. Today virtually all these firms are making significant investments in new technology solutions capable of expanding affordable Internet access as a core strategy to expand their market base.

Launch of Internet of Things (IoT)

While the past focus was largely on connecting people, late in this 2000-2015 timeframe, there has been an expanding focus on connecting things—it's no longer just people. This includes capturing more data, expanded monitoring and control systems, "big data", etc. The list is endless. Hopefully this too will have a positive spillover impact on connecting the remaining 4+ billion people, and not wind up diverting attention from the connecting people gap that still needs much attention.

Refining ICT4D to Digital Development

During the lifespan of the MDGs, there was growing recognition that an expanded "all-things-digital" orientation was needed. Hence, the migration from a focus on "ICT for Development" to a broader "Digital Development," and more recently, "Digitization", started to take root in some quarters.

Formal Recognition within SDGs (for 2015-2030) and Targets

In the MDGs created for 2000-2015, there was scant mention of technology, though the ITU, through the WSIS, moved the agenda forward in the international development community. With the SDGs, there is a formally stated recognition on the need to expand affordable Internet access. While not a high priority, there is a stake in the ground around which the international development community can focus its collective efforts. The next Chapter pursues this topic.

...experiences and observations from the author

I would like to put forward some thought and caution on a few topics:

Need to Rethink Development Projects—Many USAID administrators have put forward the notion that the role of U.S. foreign assistance is to work itself out of a job. A noble goal. But to achieve this end, there must be a reorientation in the way USAID structures and spends its currency—that currency being "projects." If projects cannot be constructed such that at their conclusion there is no longer the need for further engagement, then the role of foreign assistance will continue indefinitely—project-after-project. In the area of Digitization, there is the very real potential for bringing assistance to a logical end. In virtually every engagement I've been involved with, when the project concluded, USAID was able to exit its engagement with minimal or no adverse impact—with the trajectory for change initiated through the project, not only maintained but even expanded.

Monitoring and Evaluation—While I suspect most readers support M&E, it is important that it be used towards positive ends. That is, identifying in real-time what isn't working, making needed adjustments as the project unfolds and making improvements. M&E should add value towards our ability to achieve success within the current project. Not as a postmortem. It should not be a paper exercise, as is too often the case. I recall in the Vietnam LMI project that USAID chose the project as one to have an independent M&E analysis. Our key project focus was operationalizing a newly approved USF. We worked on this for about a year, and as the project was ending, funds were collected but not yet disbursed. Disbursements were scheduled to occur a few months later. The formal M&E could not and would not recognize the project as a success in their evaluation—after all, disbursements would not occur until after the contract performance period. It was outside the scope. Our USF project cost USAID around US$ 100K, yet the initial USF disbursement about four months later was on the order of US$ 45M. The final M&E report reflected the first disbursement as a footnote, and not as a major success of the Vietnam LMI. My copy of the M&E report ended up in the trash can, never read. No value to this project or any that followed.

Unanticipated Spillover—There is another reality often missed in many initiatives and that is the spillover impact. I recall years back taking a couple projects I had managed and creating a mind map of all the interconnecting threads—the subsequent initiatives that were directly and indirectly impacted by these two projects. Clearly these were outside the scopes of these projects. No issue there. But I was amazed at the undocumented linkages and value-add! A thought: when we think of possible successes, we need to think beyond our typical project boundaries. Success and ultimate impacts will most likely reach well beyond the project scope and influence the results achieved by others—in other projects where we may not even be aware they exist. Success should be redefined to encompass the spinoff impacts that will be realized in the future. Hard to assess, but a critical component of international development.

...experiences and observations from the author

One of the significant dynamics taking place during the timeframe of the MDGs, and currently accelerating, is the expanding role of the international high-tech firms.

Early on the most common partnership orientation was where a development agency, such as USAID, would put together some thoughts via country assessments or through the actual design of a program, and as part of the process, identify where there was an opportunity for value-add via a private sector partnership. And based on that potential, the project team would approach firms such as Cisco, Intel, Microsoft, Qualcomm, etc. And this worked!

I recall the situation in Vietnam that included a rural broadband project. I approached Intel and we partnered on two WiMAX deployments in the Lao Cai Province. After these were up and running, one of Intel's VPs who managed their SE Asia portfolio visited these deployments. He later contacted me and wanted to meet. His opening comment at our meeting was very straight forward—Intel had a list of around 20 SE Asia countries in which they wanted to deploy WiMAX—Vietnam was near the bottom of their list. It was the first to deliver a deployment. Their assessment was that the partnership with USAID was the trigger. He asked if USAID would like to partner in more countries with Intel.

More recently there are signs that the tide is shifting to where the interest goes both directions. At the start of the GBI program we became engaged with Intel—the GBI joining them in a series of workshops they were just launching. Later we became engaged with a new emerging project, Mawingu—a TV-White Space initiative in Nanyuki, Kenya, in partnership with Microsoft. The project was already underway. We had the opportunity to support this initiative to expand rural broadband. Around the same timeframe we ran across a situation in Jamaica where there was a USAID-run education initiative where extending broadband to rural schools presented an opportunity. We approached Microsoft and established a partnership around this initiative, along with a local MNO, FLOW.

In the Philippines a USAID sub-contractor (SSG-Advisors) partnered with Microsoft on a TVWS deployment for supporting a USAID-led ECOFISH initiative. In Indonesia USAID partnered with Microsoft and a local implementer for a TVWS deployment. The Indonesia project was to help shape future deployments funded by their re-packaged Universal Service Obligation (USO) Fund being supported by USAID.

In looking forward, there is an unfolding dynamic where private-sector firms are increasingly triggering development-related initiatives. The potential is significant where the high-tech industry is positioned to invest in infrastructure build-out, and where development agencies such as USAID can provide value-added support for legal and regulatory changes. These agencies can also expand support for delivering sector-specific socioeconomic value by linking to these donor support sector programs (education, health, etc.) There is the need for further focus in this space.

CHAPTER 3

Future Direction: 2015 - 2030

During the timeframe of 2013-2014, as the MDGs came towards closure, the international community once again focused on establishing not only a new set of goals and targets, but also a series of related initiatives. The focal point for this refreshed agenda was set by the United Nations through the Sustainable Development Goals (SDGs), and followed by the broader international develop community with initiatives.

Sustainable Development Goals (SDGs): 2015 – 2030

1. End Poverty In All Its Forms Everywhere
2. End Hunger, Achieve Food Security and Improve Nutrition and Promote Sustainable Agriculture
3. Ensure Healthy Lives and Promote Well-Being For All at All Ages
4. Insure Inclusive and Equitable Quality Education and Promote Lifelong Learning Opportunities For All
5. Achieve Gender Equality and Empower All Women and Girls
6. Ensure Availability and Sustainable Management of Water and Sanitation For All
7. Ensure Access to Affordable, Reliable, Sustainable and Modern Energy For All
8. Promote Sustained, Inclusive and Sustainable Economic Growth, Full and Productive Employment, and Decent Work For All
9. Build Resilient Infrastructure, Promote Inclusive and Sustainable Industrialization and Foster Innovation
10. Reduce Inequality Within and Among Countries
11. Make Cities and Human Settlements Inclusive, Safe, Resilient and Sustainable
12. Ensure Sustainable Consumption and Production Patterns
13. Take Urgent Action to Combat Climate Change and Its Impacts
14. Conserve and Sustainably Use The Oceans, Seas and Marine Resources For Sustainable Development
15. Protect, Restore, and Promote Sustainable Use of Terrestrial Ecosystems, Sustainably Manage Forests, Combat Desertification, and Halt and Reverse Land Degradation and Halt Biodiversity Loss
16. Promote Peaceful and Inclusive Societies for Sustainable Development, Provide Access to Justice For All and Build Effective and Accountable and Inclusive Institutions At All Levels
17. Strengthen The Means of Implementation and Revitalize The Global Partnership for Sustainable Development

Sustainable Development Goals (SDGs) – 2015-2030

These seventeen SDGs built off and replaced the eight MDGs, and are supported by 169 SDG Targets[16]

Leading up to the SDGs, the role of Information and Communication Technology (ICT) became increasingly recognized by the international development community as an important socioeconomic point-of-leverage.

While none of the 17 SDGs made specific reference to ICTs, a few of the targets did refer to ICTs and technology. In addition, there was a broad statement that declared:

> *"The spread of information and communication technology and global interconnectedness has great potential to accelerate human progress, to bridge the digital divide and to develop knowledge societies".*[17]

For the first time, the UN's global goals included a specific reference to leveraging ICTs, with expanding affordable Internet access reflected as one of the SDG Targets.

> **SDG Target 9c:**
> **"Significantly increase access to information and communication technology,**
> **...and strive to provide access to the Internet in least developed countries by 2020"**

Also, with the new SDGs, the UN, through an Expert Group on SDGs (IAEG-SDG), developed a set of indicators linked to several SDG Targets that will be used to track progress throughout the life of the SDGs.

SDG-Related Target Indicators for ICTs

- **Target 4a:** Proportion of schools with access to the Internet for pedagogical purposes
- **Target 4a:** Proportion of schools with access to computers for pedagogical purposes
- **Target 4.4:** Proportion of youth/adults with ICT skills, by type of skills
- **Target 5b:** Proportion of individuals who own a mobile telephone, by sex
- **Target 9c:** Percentage of the population covered by a mobile network, broken down by technology
- **Target 17.6:** Fixed Internet broadband subscriptions, broken down by speed
- **Target 17.8:** Proportion of individuals using the Internet (ITU)

---

[16] http://www.un.org/ga/search/view_doc.asp?symbol=A/RES/70/1&Lang=E
[17] http://www.itu.int/en/ITU-D/Statistics/Pages/intlcoop/sdgs/default.aspx

The challenges that the SDG Target 9c seeks to address are fundamental issues that: 1) throughout the life of the MDGs, the LDCs continued to reflect low Internet growth, and 2) the gap between the LDCs and the developed, even other developing countries, continued to increase. The data shows the gap between the LDCs and others is projected to continue throughout the life of the SDGs (2030).

Parallel with SDG recognition of ICT, the broader international development community, including the International Telecommunications Union (ITU), the World Economic Forum (WEF), the U.S. State Department, and others, have launched new, or strengthened existing, broadband-related initiatives. These initiatives most often focus on connecting the "next billion," or the "next 1.5 billion." Not the bottom billion.

These initiatives focus primarily on the first-half of SDG Target 9, *"Significantly increase access to information and communication technology."* This book focuses on the second-half of the SDG Target 9c, *"and strive to provide access to the Internet in least developed countries by 2020."*

World Development Report 2016: Digital Dividends:

Concurrent with the development, refinement, and finalization of the new SDGs, the World Bank Group published the World Development Report for 2016, with a targeted focus on Digital Dividends[18]. This report recognized the value to social and economic development through digital technologies, along with the current and expanding gap between the haves and have-nots.

In the Forward to the report, the World Bank president, Jim Yong Kim, states the following:

> *"Those in extreme poverty have the most to gain from better communications and access to information. Nearly 6 billion people do not have high-speed Internet, making them unable to fully participate in the digital economy. To deliver universal digital access, we must invest in infrastructure and pursue reforms that bring greater competition to telecommunications markets, promote public-private partnerships, and yield effective regulation. The Report concludes that the full benefits of the information and communications transformation will not be realized unless countries continue to improve their business climate, invest in people's education and health, and promote good governance. In countries where these fundamentals are weak, digital technologies have not boosted productivity or reduced inequity. Countries that complement technology investments with broader economic reforms reap digital dividends in the form of faster growth, more jobs, and better services."*

---

[18] http://documents.worldbank.org/curated/en/896971468194972881/pdf/102725-PUB-Replacement-PUBLIC.pdf

The report makes a strong case for digital transformations that include connecting people, businesses, and governments. It goes on to state:

> *"The lives of the majority of the world's people remain largely untouched by the digital revolution. Only around 15% can afford access to broadband Internet. Mobile phones, reaching almost four-fifths of the world people, provide the main form of Internet access in the developing world. But even then, nearly 2 billion people do not own a mobile phone, and nearly 60 percent of the world's population has no access to the Internet. The world's offline population is mainly in India and China, but more than 120 million people are still offline in North America. The digital divide within countries can be as high as that between countries...."*

The report documents how the Internet promotes development by promoting inclusion, efficiency, and innovation, with the digital dividends being growth, jobs and service delivery, along with leading to more trade, better capital use, and greater competition. The report provides excellent insights into what it refers to as "Digital Enablers," with rich research, analytics, and examples.

Broadband Commission for Sustainable Development

Established in 2010 by the ITU and UNESCO, the Broadband Commission promotes effective, inclusive, and sustainable development of the broadband, including technologies, policy and regulations, broadband for development and policy recommendations. The Commission offers strong advocacy and support/guidance in the areas of building National Broadband Plans. Its focus is on imbedding the expansion and adoption of Broadband in supporting the SDGs and the SDG Targets.

Towards this priority, the Commission periodically publishes annual reports. The State of Broadband 2017[19] is their most recent (published in September 2017). In announcing this new report from the Broadband Commission in New York on 18 September 2017, the UN Secretary-General, Antonio Guterres made the following statement: [20]

> *"The membership of this Commission offers an encouraging example of just the kind of multi-stakeholder partnerships we need to achieve the Sustainable Development Goals. Technology is crucial in empowering people to participate in our digital future, and in helping governments to better serve people. But we must also address significant concerns such as cybersecurity, human rights, privacy, as well as the digital divide, including its gender dimensions. Broadband is a remarkable tool; now we must do more to ensure that all enjoy its benefits. Developing countries face the very real risk of being left behind. I look to this*

---

[19] https://www.itu.int/dms_pub/itu-s/opb/pol/S-POL-BROADBAND.18-2017-PDF-E.pdf
[20] http://www.itu.int/en/mediacentre/Pages/2017-PR47.aspx

Commission to help ensure that broadband charts a course that includes all humankind, enhances human dignity and serves the global good."

At this same session, Houlin Zhao, ITU Secretary-General, who serves as co-Vice Chair of the Commission stated:

"This year's State of Broadband 2017 report highlights several important findings. First, there is a suggestion that we are entering a 'winner takes all' phase in digital development – digital 'frontrunner' countries are moving even further ahead, while developing countries are generally being left behind. Furthermore, gaps in transmission speeds are also increasing. And there is still no visible progress that the digital gender divide is closing. Even in a high-growth industry such as ours, there is still cause for concern. In a few weeks, I'll address the seventh World Telecommunication Development Conference (WTDC) in Buenos Aires. My message will be clear: It is our responsibility to bring the power of ICTs to all nations, all people and all segments of society."

And Irina Bokova, UNESCO Director-General who also serves as co-Vice Chair, expanded off her earlier statement on this topic and said:

"Today, more than ever, the digital revolution must be a development revolution ... a sustainable development revolution. We need broadband to strengthen the sustainability of development efforts. We need broadband to bridge divides and not deepen them – especially for girls and women. We need broadband that ensures equal access to education, that enhances the quality of learning across the world, because these are the strongest foundations for sustainability and peace."

It is clear that broadband and its adoption is an uppermost priority at the highest levels of the United Nations, reflected by the members of the Broadband Commission for Sustainable Development. The foundation is established. The focus must now shift to the action!

International Telecommunications Union: Connect 2020

ITU's Connect 2020[21] was approved at the ITU Plenipotentiary Conference (PP-14) with a series of goals and targets that member states have committed to achieve by 2020—a date consistent with SDG Target 9c. In many ways Connect 2020 builds off of, melds, and advances the earlier WSIS Targets and the Broadband Commission Targets for 2015. These fall into four Goals, with refined Targets within each of these Goals.

ITU Goals and Targets for 2020

Goal 1: Enable and foster access to and increased use of Telecommunications/ICTs
  Target 1.1: Worldwide, 55% of households should have access to the Internet
  Target 1.2: Worldwide, 60% of individuals should be using the Internet
  Target 1.3: Worldwide, telecommunications/ICT should be 40% more affordable

Goal 2: Inclusiveness - Bridge the digital divide and provide broadband for all
  Target 2.1A: In the developing world, 50% of households should have access to the Internet
  Target 2.1B: In the LDCs, 15% of households should have access to the Internet
  Target 2.2A: In the developing world, 50% of individuals should be using the Internet
  Target 2.2B: In the LDCs, 20% of individuals should be using the Internet
  Target 2.3A: The affordability gap between developed and developing countries should be reduced by 40%
  Target 2.3B: Broadband services should cost no more than 5% of average monthly income in developing countries
  Target 2.4: Worldwide, 90% of the rural population should be covered by broadband services
  Target 2.5A: Gender equality among Internet users should be reached
  Target 2.5B: Enabling environments ensuring accessible telecommunications/ICTs for persons with disabilities should be established in all countries

Goal 3: Sustainability – Manage challenges resulting from telecommunications/ICT development
  Target 3.1: Cybersecurity readiness should be improved by 40%
  Target 3.2: Volume of redundant e-waste to be reduced by 50%
  Target 3.3: Greenhouse gas emissions generated by telecommunications/ICT sector to be decreased per device by 30%

Goal 4: Innovative and partnership – Lead, shape and adapt to the changing telecommunication/ICT environment
  Target 4.1: Telecommunications/ICT environment conducive to innovation
  Target 4.2: Effective partnerships of stakeholders in the telecommunications/ICT environment

---

[21] http://www.itu.int/en/connect2020/Pages/default.aspx

Emerging Multilateral and Bilateral Initiatives

The above reflects a dynamic relative to the international development community placing a higher priority on ICT4D/Digital Development. While this theme emerged during the timeframe of the MDGs through the ITU/WSIS, the Broadband Commission and CSTD, it is now more pronounced with the formalization of the new SDGs covering the next 15 years (2015-2030).

The following reflects a range of new initiatives emerging in the 2015+ timeframe. Collectively these reflect new or refined ICT-related programs by key players within the international development community.

World Economic Forum: Internet for All

The WEF's Internet for All[22] is a subset of WEF's Future of the Internet Global Challenge initiative. The Future of the Internet seeks to strengthen effectiveness and cooperation of Internet governance where participants will:

- Focus on multiple projects, task forces and continuous opportunities to collaborate throughout the year on such issues as cybercrime and Internet access

- Engage with the public and private sectors to ensure perspectives are balanced across all stakeholder groups

- Contribute insights through dedicated briefings, calls and through annual updates, and

- Engage in impact-oriented discussions and activities at the regional and country level

The Internet for All specifically seeks to develop and catalyze new models of public-private collaboration with the aim of increasing affordable Internet access and relevant adoption. It is action oriented with an objective to accelerate the achievement of Internet for all, in partnership with the governments involved and under their leadership, and will involve private sector, civil society, experts, and the multilateral/bilateral community. The models developed in the country programs will serve as a basis for scaling up and replicating "internet for all" in

---

[22] https://www.weforum.org/projects/internet-for-all

countries and regions around the world. Its goal is to develop a scalable, replicable model for public-private collaboration that accelerates Internet adoption for the 4 billion people currently not on the Internet.

The project goals are structured in two phases:

- **Internet for All report (2015):** Successful practices on internet access and internet adoption initiatives to connect the unconnected;

- **Country programs (2016 onwards):** The first country program, with the full endorsement already provided by the government Ministers involved, was identified, subject to final approval at the Annual Meeting 2016, in the Northern Corridor countries of Rwanda, Kenya, Uganda, and South Sudan. Additional country programs (up to three in total) in other regions of the world (Asia and Latin America) were also scoped in 2016, with India already identified as a potential 2nd country program.

U.S. Congress: Digital GAP Act

In early 2017, Congress passed the Digital Global Access Policy Act or Digital GAP Act[23]. This bill states that it is U.S. policy to coordinate with foreign governments, international and regional organizations, businesses, and civil society to close the digital gap in developing countries.

Section 6 of the GAP Act states that:

(1) The Department of State should seek to enhance the effectiveness of U.S. foreign assistance efforts in carrying out the policies and objectives of this bill, including re-designating an existing Assistant Secretary position in the State Department to be the Assistant Secretary for Cyberspace; and

(2) the U.S. Agency for International Development should integrate efforts to expand Internet access and establish guidelines for the protection of personal information of individuals served by humanitarian, disaster, and development programs.

As this book is being written, it is too early to determine the actual initiatives and results coming out of this Act relative to budget levels, subsequent program designs, implementations, and results.

---

[23] https://www.congress.gov/bill/115th-congress/house-bill/600

U.S. State Department: Global Connect Initiative (GCI)

The State Department launched the GCI[24] in 2015. It focuses on expanding access to the Internet by 1.5 billion in developing countries by 2020—a target in support of the SDG Target 9c.

The objectives of the GCI include:

1. Work with every stakeholder group to mainstream the view that Internet connectivity is as fundamental to economic development as roads, ports and electricity;

2. All countries integrate Internet connectivity and digital technologies as part of national development strategies;

3. International Development institutions, such as multilateral development banks and development agencies, prioritize Internet for Development; and

4. Catalyze and support innovative industry driven solutions.

Other priority areas of the GCI include:

1. OPIC

    - $250 Million in financing for development of a network of 2,500 telecom towers across Burma

    - Similar efforts across Kenya with a local provider of solar-powered wireless Internet

2. USAID, FCC and MCC

    - The FCC Chairman, USAID's Acting Administrator, and MCC's CEO, affirmed commitments to extend connectivity through their programs at the 9/27/2017 launch of the Global Connect Initiative.

    - Work with all development organizations and produce strategies for helping to bridge the digital divide in key countries and regions in advance of April 2016 Conference

---

[24] https://blogs.state.gov/stories/2016/01/25/global-connect-initiative-making-internet-development-priority

At the time this book was written, it was unclear as to whether the GCI will continue under the Trump administration and if so, what will be the priority areas of focus. The initial input is encouraging, but I am unaware as to whether any final decisions have been made.

Digital Impact Alliance (DIAL)

DIAL[25] is a relatively new initiative put into place with support from the UN Foundation, USAID, Sida and the Bill and Melinda Gates Foundation (BMGF). DIAL's vision is to:

> *"Realize a more inclusive digital economy for the underserved in emerging markets, whereby all women, men and children benefit from life-enabling, mobile-based digital services."* Its mission is to: *"accelerate the collective efforts of government, industry and development organizations to realize this vision."*

DIAL envisions the opportunity of the digital revolution that holds tremendous promise for advancing GDP growth, yet recognizes that gaps exist. To realize this opportunity, barriers such as fragmentation, expertise, shared understanding, and achieving value need addressing.

An initial focus of DIAL is currently on what is referred to as "Principles for Digital Development."[26]

USAID: Digital Inclusion

With the ending of the GBI program, USAID's focus on connectivity became part of a larger Digital Development theme.

The following was extracted from a recent USAID Center for Digital Development publication[27]:

> *"The Digital Inclusion team helps bridge the digital divide by expanding access to the Internet in countries where USAID works to accelerate the Agency's development objectives and ensuring the most marginalized have the skills and resources to be active participations in the digital economy. Through public- private partnerships and direct technical assistance the team has unlocked more than $100 million for expanded internet access, resulting in approximately 20 million new mobile and internet users worldwide, and directly supported USAID health and education programs by connecting 56 schools and 39 hospitals to the Internet in 2013."*

---

[25] http://digitalimpactalliance.org
[26] https://digitalprinciples.org
[27] https://www.usaid.gov/digital-development

Their Connectivity and Access for Development materials further states:

"The incredible proliferation of mobile phones and broadband Internet offers a profound opportunity to connect people around the world and can increase effectiveness, efficiency, and connectedness across sectors, in health, agriculture, education, and governance. According to one estimate, the "digitization" of developing economies could yield as much as $4.1 trillion in GDP among the billion consumers at the base of the pyramid. This would create 64 million new jobs and help lift 580 million people out of poverty who currently live on less than US$4.00 per day. Yet significant barriers still persist for people in developing countries in effectively using digital technology, particularly for women and rural poor. These include affordability, illiteracy and the lack of locally relevant content, inadequate digital skills, and a lack of demographic usage and access data.

The approach focuses on what the Center for Digital Development refers to as *"Connected Programs,"* and *"Enabling Environment Investments."*

During late July 2017, USAID's Center for Digital Development (CDD) issued a Request for Application (RFA). The scope of the RFA reached beyond the DI components to include a broader range of technology support provided through the Lab's Center for Digital Development. This RFA sought to put into place a Grant with the following Background and Objective put forward in the RFA:

Background: *"The Center for Digital Development (CDD), which leads the Lab's 'Technology' objective, has as its mission to ensure that digital economies are inclusive and sustainable, and that they improve the lives of millions of poor and vulnerable people throughout the developing world. As importantly, CDD is working to ensure that USAID leverages and mainstreams digital tools and the adaptive, data-driven approaches these tools enable to transform how development is done. CDD is achieving this mission by prioritizing technical areas of work through four teams : (1) the Digital Inclusion team is helping to bridge the digital divide by expanding access to the internet in countries where USAID works to ensure the most marginalized have the skills and resources to be active participants in the digital economy, (2) the Development Informatics team seeks to make development more adaptive, efficient, and responsive to citizens and decision makers by helping transform the use of data and technology throughout development, (3) the GeoCenter Plus team applies geographic analysis to international development challenges to improve the strategic planning, design, monitoring, and evaluation of USAID's programs, and (4) the Digital Finance team is acting on the growth in digital financial services to help the world's financially excluded and underserved populations access the use of financial services that meet their needs. These four technical areas are making progress towards their respective objectives by 1) building the Agency's technical capacity in the use of digital technologies and data driven approaches, and 2) supporting public goods, partnerships, and policy change that improve the enabling environment for an inclusive digital economy."*

Objective: *"The objective of this award will aim to increase the use of and access to digital financial services, mobile and internet, and increase the use of data for strategic planning and adaptive programming within USAID missions, bureaus, and independent offices. This award will work with organizations to conduct diagnostic services, program design and implementation, research projects, and develop tools and approaches that will help reach this objective. The prime recipient of this award will receive technical direction from the Agreement Officer's Representative (AOR) and technical managers within USAID to guide their work activities."*

In September 2017, an award was made to Development Alternatives, Incorporated (DAI) for what is referred to as the **"Digital Frontiers"** (DF) project. This grant is for the years 2017-2022, for a total estimated amount of between US$ 60-75M.

Private Sector Initiatives

The private high-tech sector has led the way in developing ever-changing higher-capacity, lower-cost solutions for the marketplace. This dynamic started with costly mainframe computing resources. Over the last less-than 50 years these technology advances put what used to be mainframe computer power into small hand-held devices. With this, usable "Apps" that require minimal high-tech skills at the user-level were added. Further, the technology advancement has delivered affordable wireless broadband connectivity such that these small personal devices have near-constant broadband access to-from a rapidly expanding array of cloud-services that add even greater functionality and value.

For most high-tech firms such as Google, Facebook, Microsoft, Cisco, and others, this accelerating dynamic of providing cloud-based services has already become core to their corporate strategy and is the essential component for their future growth. These firms are making considerable investment towards advancing connectivity technologies along with expanding their deployment. The focal point is to accelerate adoption, allowing these firms to enter new markets that support their corporate growth strategies. These dynamics create a rich opportunity for building partnerships with the international development community, be it at the multilateral or the bilateral levels.

The following represents a sampling of activities these firms are currently engaged in. These hold potential for expanding broadband connectivity into the Developing, and even the Least Developed Countries:

- **Google**: Google became engaged in fiber through their Google Link projects[28], including several existing and planned deployments in Africa. Google is also engaged in a sophisticated drone program for delivering wireless broadband. The initiative is called SkyBender[29], with tests currently taking place using prototype 5G technologies. Google is also refining its development of high altitude balloons for providing Internet connectivity, through its Loon[30] initiative.

---

[28] https://www.google.com/get/projectlink/

[29] https://www.theguardian.com/technology/2016/jan/29/project-skybender-google-drone-tests-internet-spaceport-virgin-galactichttps://www.theguardian.com/technology/2016/jan/29/project-skybender-google-drone-tests-internet-spaceport-virgin-galactic

[30] http://www.theregister.co.uk/2016/05/19/project_loon_googles_global_internet_megaplan/

- **Facebook**: Facebook has both a fiber-based program[31] as well as a drone program[32] with OpenCellular[33] under development to expand global Internet access.

- **Microsoft**: Microsoft is actively engaged in fiber (see above-referenced undersea initiative in partnership with Facebook), but has also been a champion of terrestrial TV White Space[34] (TVWS) technology development and deployments. As an example of partnership of a high-tech firm with the international development community, USAID partnered with Microsoft on several TVWS deployments in Kenya, Jamaica, Botswana, Philippines and Indonesia.

- **SpaceX, Boeing and OneWeb**: The media often focuses SpaceX[35] attention on its reusable launch crafts. But behind this key component of their strategy is the development and launch of a satellite constellation of 4,000 communication satellites in 1,100 km altitude orbits for providing global Internet services. They are hoping to start operations in 2020. Boeing[36] too has entered this space, having bought out Hughes Space and Communications in 2000. OneWeb[37] is another contender with plans underway to launch a constellation of 648 satellites.

...experiences and observations from the author

A growing opportunity likely to emerge during the timeframe of the SDGs will come from developing an even closer relationship between the donor agencies and the private sector firms that are increasingly engaged, and making investments in this space.

Experience has shown tremendous synergy when USAID focused its resources on improving legal and regulatory issues. This creates an enabling environment for the private sector. This type of public-private partnership advances expanding market entry opportunities as well as expanding current in-country business opportunities.

The international high-tech private sector firms most often partner with local firms in expanding deployments, capacity development, hosting of locally relevant content, etc. From my experience, these are rich partnerships, with the future providing game-changing opportunity for taking these even further.

---

[31] http://www.wsj.com/articles/facebook-and-microsoft-to-build-fiber-optic-cable-across-atlantic-1464298853
[32] http://www.theverge.com/a/mark-zuckerberg-future-of-facebook/aquila-drone-internet
[33] https://www.wired.com/2016/07/facebook-2/
[34] https://www.microsoft.com/en-us/research/project/dynamic-spectrum-and-tv-white-spaces/
[35] https://en.wikipedia.org/wiki/SpaceX
[36] http://www.reuters.com/article/space-satellite-ula/lockheed-boeing-rocket-venture-to-launch-micro-satellites-idUSL1N13E24920151119
[37] http://oneweb.world

CHAPTER 4

LDCs:
The Ultimate Challenge

Least Developed Countries (LDCs)

The two core areas of focus within this book are "Digitization" and the Least Developed Countries (LDCs)—those countries where there is a lag in Internet build-out and adoption. The LDC focus is very simple in that:

1. There is the SDG Target 9c that focuses specifically on expanding affordable Internet in the LDCs;

2. The vast majority of recent initiatives, launched in part based on the SDGs, are not focusing on the LDCs. Rather, their focus is on countries where the market-place dynamics will most likely resolve the bulk of connectivity issues without much, if any direct assistance;

3. As a percent of the country population, those living in extreme poverty and in rural areas are the highest in the LDCs; and

4. The LDCs are countries where the current business models have had, and are currently having, minimal success and impact.

If success is to be achieved in the LDCs, there is simply the need for an updated, more innovative approach. Collectively the LDCs represent the ultimate challenge. And if successful business and technology can be found here, then other developing countries will also benefit.

The UN currently classifies 48 countries as Least Developed Countries (LDCs). Most of these countries are relatively small in geographic size and population with a total LDC population of just under 1 billion.

The average population of an LDC country is under 20 million, though many are significantly less. The LDC total population is also predominantly rural, with an average of 68% living in rural communities. And 72% of the LDC population lives at or below the extreme poverty level of US$ 1.90/day. With regards to access to mobile, the average subscription rate is placed at 69% of the population. However, with regards to individual Internet use, the average is on the order of just 18%.

With regards to Internet access, the following graphic reflects the most recent 2017 data from the ITU's Broadband Commission for Sustainable Development report, "The State of the Broadband 2017"[38]. This data is at the household level by geographic region and level of development.

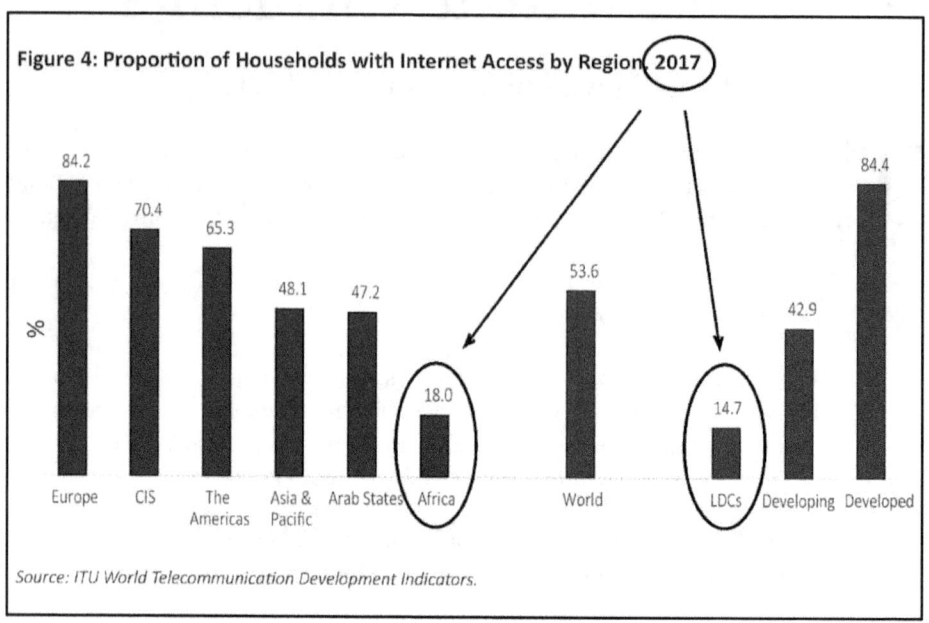

As a general statement, two key elements are often missing in the LDCs: 1) lack of high-level public and private sector commitment to address this issue as a national priority, and 2) substantially less individual and organizational (public and private sector) market demand than what exists in more densely populated, more mature developing economies, and where a larger percent of the population live in urban areas. These two elements require a comprehensive and innovative approach to achieve success.

Further, for success to occur, the focus cannot be placed only on expanding connectivity. Rather, it must also include placing a priority on the broader ecosystem of ICT—yes, expanding affordable Internet where it doesn't exist in rural areas; but there is also the need to address and leverage all-things-digital. "Digitization."

---

[38] https://www.itu.int/dms_pub/itu-s/opb/pol/S-POL-BROADBAND.18-2017-PDF-E.pdf

...experiences and observations from the author

Picking up on the rural poor theme, I am reminded of an incident that took place in May 2007. Intel had invited me to speak at a WiMAX Forum in Taipei, Taiwan. Approximately 20 Asia-Pacific countries were represented at the event. My presentation was about an early deployment of two broadband wireless networks in rural Vietnam. These deployments were undertaken in parallel to, and supportive of, operationalizing the Vietnam Telecommunications Fund (VTF).

As I ended my presentation, I challenged the approximate 400 public and private sector participants to adopt a view that the lack of advancing telecommunications to those living in rural areas, where most often the poorest populations live, be thought of as a form of violence. I shared with the attendees a quote from an unknown author that I heard 30 years earlier...a quote that has driven my passion, *"No one should suffer the violence of being ignored."* My concluding message was that collectively we should place a priority on solving this critical issue. The emerging technologies discussed at the Forum demonstrated that we were quickly moving towards technology solutions that can address this issue.

At the closing luncheon, I had the opportunity to sit with and discuss this topic with the Forum organizers. In giving the closing speech, the Chairman of the Forum referenced back to my theme and reinforced it as a priority – the priority need for a rural focus by those regional governments and companies in attendance.

10 years later, this critical need remains the most prevalent situation today in many countries. And unfortunately, it will likely be with us for the next 10 years and beyond, unless the international community places a high priority on this issue and follows-up with action.

No one should suffer the violence of being ignored

Author Unknown

LDC and LCC Country Profiles

The following pages reflect a series of regional-geographic tables that provide country-level profile information for each of the 48 Least Developed Countries (LDCs). As noted in the tables, the UN/ITU classified a number of these LDCs as being Least Connected Countries (LCCs). A separate set of tables towards the end of this chapter reflects those 8 countries that are classified as LCCs but are not classified as LDCs.

The data in the following tables is at the Individual Use level. Following is a quick summary of what to consider when reviewing Individual User access for the LDCs.

- The average population of an LDC is under 20 million, though many are significantly less.

- The total population of these LDCs is also predominantly rural, having an average of 68% living in rural communities.

- Significant portions, about 72% of LDC population, live in extreme poverty—making less than $1.90 per day.

- With regards to access to mobile telephone, the average subscription rate is placed at 69% of the population.

- The individual Internet Use is on the order of 18%, just slightly over the 14.7% reflected at household level.

Least Developed Countries (LDCs)

Africa LDCs: (1 of 2)					
Country	Total Population 2015[1]	Poverty Population[2] Population % Pop	LDC[3] LCC[4]	Rural Population[5] Population % Pop	Access Statistics 2015 %Mobile[6] %Internet[7]
Angola	25.0M	7.5M 30.1%	X X	14.0M 56.0%	60.84 12.40
Benin	10.9M	5.6M 51.6%	X X	6.1M 56.0%	85.64 6.79
Burkina Faso	18.1M	7.9M 43.7%	X X	12.7M 70.2%	80.64 11.39
Burundi	11.2M	8.7M 77.7%	X X	9.8M 87.5%	46.22 4.87
Central Africa Rep	4.9M	3.2M 66.3%	X	2.9M 59.2%	25.87 4.56
Chad	14.0M	5.4M 38.4%	X X	10.9M 77.9%	40.17 2.70
Comoros	0.8M	0.1M 13.5%	X	0.6M 75.0%	50.90 7.46
Dem Republic Congo	77.3M	59.6M 77.1%	X X	44.4M 57.4%	52.99 3.80
Djibouti	0.9M	0.2M 22.5%	X X	0.2M 22.2%	34.94 11.92
Equatorial Guinea	0.8M	0.3M 40% e	X X	0.5M 62.5%	66.72 21.32
Eritrea	6.3M	2.5M 40% e	X	1.3M 20.6%	7.05 1.09
Ethiopia	99.4M	33.3M 33.5%	X X	80.0M 80.5%	42.76 11.60
The Gambia	2.0M	0.9M 45.3%	X X	0.8M 40.0%	137.85 17.12
Guinea	12.6M	4.4M 35.3%	X X	7.9M 62.7%	87.17 4.70
Guinea-Bissau	1.8M	1.2M 67.1%	X X	0.9M 50.0%	69.27 3.54
Lesotho	2.1M	1.3M 59.7%	X X	1.6M 76.2%	100.94 16.07
Liberia	4.5M	3.1M 68.6%	X X	2.3M 51.1%	81.09 5.90
Madagascar	24.2M	18.2M 77.8%	X X	15.7M 64.9%	44.12 4.17

Least Developed Countries (LDCs)

Country	Total Population 2015	Poverty Population		LDC LCC	Rural Population		Access Statistics 2015
		Population	% Pop		Population	% Pop	%Mobile %Internet
Malawi	17.2M	12.2M	70.9%	X X	14.4M	83.7%	37.94 9.30
Mali	17.6M	8.7M	49.3%	X X	10.6M	60.2%	139.61 10.34
Mauritania	4.1M	0.3M	5.9%	X X	1.6M	39.0%	89.32 15.20
Mozambique	28.0M	19.2M	68.7%	X X	19.0M	67.9%	74.24 9.00
Niger	19.9M	9.1M	45.7%	X X	16.2M	81.4%	46.50 2.22
Rwanda	11.6M	7.0M	60.4%	X X	8.3M	71.6%	70.48 18.00
São Tomé & Principe	0.2M	0.1M	32.3%	X	0.1M	50.0%	65.09 25.82
Senegal	15.1M	5.7M	38.0%	X X	8.5M	56.3%	99.95 21.69
Sierra Leone	6.5M	3.4M	52.3%	X	3.9M	60.0%	80.53 2.50
Somalia	10.8M	4.3M	40% e	X	6.5M	60.2%	52.47 1.76
South Sudan	12.3M	5.3M	42.7%	X X	10.0M	81.3%	24.50 15.90
Sudan	40.2M	7.0M	14.9%	X X	26.6M	66.2%	70.53 26.61
Tanzania	53.5M	24.9M	46.6%	X X	36.6M	68.4%	75.86 5.36
Togo	7.3M	4.0M	54.2%	X X	4.4M	60.3%	67.71 7.12
Uganda	39.0M	13.5M	34.6%	X X	32.8M	84.1%	52.43 19.22
Zambia	16.2M	10.4M	64.4%	X X	9.6M	59.3%	84.79 21.00
LDC Africa Summary	616.4M	298.5M	48.4%	34 28	421.7M	68.4%	61.13 12.68

Least Developed Countries (LDCs)

Asia LDCs: (1 of 1)					
Country	Total Population 2015	Poverty Population	LDC LCC	Rural Population	Access Statistics 2015
		Population % Pop		Population % Pop	%Mobile %Internet
Afghanistan	32.5M	13.0M	X	23.8M	61.58
		40%e	X	73.2%	8.26
Bangladesh	161.0M	29.8M	X	105.8M	81.90
		18.5%	X	65.7%	14.40
Bhutan	0.8M	0.02M	X	0.5M	92.18
		2.2%		62.5%	39.80
Cambodia	15.6M	0.3M	X	12.3M	133.00
		2.2%		78.8%	19.00
East Timor-Leste	1.2M	0.6M	X	.8M	117.40
		46.8%		66.7%	13.40
Laos PDR	6.8M	1.14M	X	4.2M	53.10
		16.7%	X	61.8%	18.20
Myanmar	53.9M	21.6M	X	35.6M	75.68
		40%e	X	66.0%	21.80
Nepal	28.5M	4.3M	X	23.2M	96.75
		15.0%	X	81.4%	17.58
Yemen	26.8M	10.7M	X	17.5M	67.98
		40%e	X	65.3%	25.10
LDC Asia Summary	327.1M	81.46M	9	223.7M	81.00
		25.0%	6	68.4%	16.50

Least Developed Countries (LDCs)

Pacific Islands LDCs: (1 of 1)

Country	Total Population 2015	Poverty Population		LDC LCC	Rural Population		Access Statistics 2015	
		Population	% Pop		Population	% Pop	%Mobile	%Internet
Kiribati	0.1M	0.01M	14.1%	X X	0.06M	60.0%	38.84	13.00
Solomon Islands	0.6M	0.3M	45.6%	X X	0.5M	83.3%	72.66	10.00
Tuvalu	0.01M	0.0M	2.7%	X	0.004M	40.0%	40.34	42.70
Vanuatu	0.3M	0.1M	15.4%	X	0.2M	74.0%	66.25	22.35
Pacific Islands Summary	1.01M	0.04M	40.6%	4 2	.764M	75.6%	68.68	14.14

Latin America LDCs: (1 of 1)

Country	Total Population 2015	Poverty Population		LDC LCC	Rural Population		Access Statistics 2015	
		Population	% Pop		Population	% Pop	%Mobile	%Internet
Haiti	10.7M	5.8M	53.9%	X	4.4M	41.1%	68.84	12.20
Latin America Summary	10.7M	5.8M	53.9%	X	4.4M	41.1%	68.84	12.20

Global-Wide LDC Summary (including only those LCCs that are also LDCs)

Country	Total Population 2015	Poverty Population		LDC LCC	Rural Population		Access Statistics 2015	
		Population	% Pop		Population	% Pop	%Mobile	%Internet
LDCs: Total	955.2M	648.9M	72.3%	48 36	650.6M	68.1%	69.10	17.70

Least Connected Countries (LCCs)

Africa					
Country	Total Population 2015	Poverty Population	LDC LCC	Rural Population	Access Statistics 2015
		Population % Pop		Population % Pop	%Mobile %Internet
Cameroon	23.3M	5.6M 24.0%	X	10.6M 45.5%	71.85 20.68
Cóte d'Ivoíre	22.7M	6.6M 29.0%	X	10.4M 45.8%	119.31 21.00
Nigeria	182.2M	97.5M 53.5%	X	95.2M 52.2%	82.19 47.44
Swaziland	1.3M	0.6M 42.0%	X	1.0M 76.9%	73.20 30.38
Zimbabwe	15.6M	3.3M 21.4%	X	10.6M 67.9%	70.00e 25.00e
LCC Africa Summary	245.1	113.6 46.4%	5	127.8 52.1%	85.12 40.93

Asia					
Country	Total Population 2015	Poverty Population	LDC LCC	Rural Population	Access Statistics 2015
		Population % Pop		Population % Pop	%Mobile %Internet
India	1,311.1M	278.0M 21.2%	X	881.2M 67.2%	78.06 26.00
Pakistan	188.9M	11.5M 6.1%	X	115.7M 61.2%	66.92 18.00
LCC Asia Summary	1,500.0M	289.5M 25.0%	2	996.9M 66.5%	76.66 24.99

Latin America					
Country	Total Population 2015	Poverty Population	LDC LCC	Rural Population	Access Statistics 2015
		Population % Pop		Population % Pop	%Mobile %Internet
Cuba	11.4M	4.4M 40%e	X	2.6M 16.5%	29.65 29.07
LCC Latin America Summary	11.4M	4.4M 40%w	1	2.6M 16.5%	29.65 29.07

Least Developed and Connected Countries (LDCs and LCCs)

Global-Wide: All LDC and LCC Summaries					
Country	Total Population 2015	Poverty Population	LDC LCC	Rural Population	Access Statistics 2015
		Population % Pop		Population % Pop	%Mobile %Internet
LDCs	955.2M	648.9M 72.3%	48 36	650.6M 68.1%	69.10 17.70
LLCs only	1,756.5M	407.5M 23.2%	8	1,127.3M 64.2%	79.26 27.29
Totals	2,711.7M	1,056.4M 39.0%	56	1,777.1M 65.6%	75.68 23.91

Note 1: 75% of the LDCs are also LCCs

---

[1] Source is WB (http://data.worldbank.org/indicator/SP.POP.TOTL) 2015 data
[2] Source is WB (https://openknowledge.worldbank.org/handle/10986/26447) – (% population below $1.90/day and total country population – date data varies per country – page 22+)
[3] Source is UN (http://www.un.org/en/development/desa/policy/cdp/ldc/ldc_list.pdf) - as of May 2016
[4] Source is ITU – Lowest 44 countries using their ICT Development Index (http://www.itu.int/en/ITU-D/Statistics/Pages/default.aspx) – 2013 data
[5] Source is WB – (http://data.worldbank.org/indicator/SP.RUR.TOTL) 2015 data (total rural population and % rural population to total population
[6] Source is ITU = Mobile-Cellular Telephone Subscriptions as a percent of the population-2000-2015 data - http://www.itu.int/en/ITU-D/Statistics/Pages/stat/default.aspx
[7] Source is ITU = Percent of Individuals using the Internet-2000-2015 data - http://www.itu.int/en/ITU-D/Statistics/Pages/stat/default.aspx

CHAPTER 5

"Digitization" and Other ICT Ecosystem Indices

Evolution of Thinking – ICT4D to DD to DtD

Over the past several years there was continuing refinement within the international development community as to the nature and breadth of Information and Communications Technologies (ICTs) as applied to Development (ICT4D). Previously there was a built-in bias towards "technology," but more recently the ICT4D orientation is giving way to a more all-things-digital focus, expressed as "Digital Development (DD)." This term is expanding in its adoption across the international development community.

This book puts forward a further refinement, "Development through Digitization" (DtD). From the author's perspective, the shifting focus is from ICT or even Digital towards an even broader "Digitization" ecosystem. It reflects a critical evolution in this space. The thinking behind DtD is twofold: 1) there is the need to make development the primary focus and digitization a program focus that contributes to development, and 2) there is the need to look beyond technology and look at the broader ecosystem of digitization that incorporates the technology, application, and adoption of services for achieving socioeconomic impact.

Digitization: Overview

In 2012, Booz&Co issued a report titled, "Digitization in Emerging Economies: Unleashing the Opportunities at the Bottom of the Pyramid."[39]

The Executive Summary opens the report with the statement shown on the right:

> "The bottom of the pyramid represents the greatest opportunity for capturing the gains in job creation and GDP growth associated with digitization, to say nothing of the salutary effect on the lives of 3.9 billion people."

---

[39] https://www.strategyand.pwc.com/media/file/Strategyand_Digitization-in-Emerging-Economies.pdf

The Executive Summary of this report elaborates even further on this theme:

> "Digitization in emerging countries could deliver as much as US$ 6.3 trillion in additional nominal GDP and 77 million new jobs over the next 10 years. Capturing this rich return will require a concerted public and private effort to bring digitization to the world's poorest people—those at the bottom of the pyramid. This population is distributed across middle- and low-income developing countries and numbers 3.9 billion in total, including 95 percent of South Asia's population, 68 percent of the Middle East and North Africa's (MENA) population, and 27 percent of Latin America's population."

Raul Katz, one of the original thought leaders in the Digitization space, has for years advanced the concept of "Digitization" through his work with several organizations, including for a period of time, Booz & Co. Mr. Katz's focus is on drilling down and quantifying the value-add achieved through Digitization.

His research puts forward that it is the adoption of all-things-digital where the sought-after economic value is ultimately derived. Connectivity is but one of the needed core elements. His research and analysis demonstrate that the derived value varies considerably industry-to-industry. Mr. Katz defined Digitization as shown in the insert.

This reorientation has profound impact with regards to the potential for deriving digitally-based value within the Least Developed Country setting.

> **Digitization:**
>
> "At the most basic level, digitization is the process of converting analog information into a digital format.
>
> In a broader context, digitization is defined as the social transformation triggered by the massive adoption of digital technologies to generate, process, share and transact information."
>
> **Raul Katz**

The "Digitization" Model

Beginning in 2012, Booz & Co, and subsequently Price Waterhouse Coopers (PWC), undertook significant bodies of research that moved beyond the topic of correlating ICT and economic impact to focus on establishing a causality relationship. The research was produced in individual reports. In recent years, some of these reports were included as chapters in the WEF's annual Global Information Technology Report (GITR). Some of the key reports presenting this research include:

- Digitization and Prosperity[40]

- Measuring Socioeconomic Digitization: A Paradigm Shift[41]

- The 2012 Industry Digitization Index[42]

- Maximizing the Impact of Digitization[43]

- Digitization for Economic Growth and Job Creation: Regional and Industry Perspectives[44]

- ICTs, Income Inequality, and Ensuring Inclusive Growth[45]

By proposing the term, "Digitization," these researchers sought to advance the thinking related to ICTs in development. But more importantly, in this book, Digitization seeks not only to incorporate a broader view of ICT (or even the digital ecosystem) with a correlation to socioeconomic development (e.g., WEF's NRI), but it also puts forward quantitatively the causality of the Digitization.

The following discussion highlights this approach, with materials extracted from one of the Booz & Co 2012 report, "Maximizing the Impact of Digitization".

---

[40] http://www.strategy-business.com/article/00127?gko=efe69
[41] http://papers.ssrn.com/sol3/papers.cfm?abstract_id=2070035
[42] http://www.strategyand.pwc.com/reports/2012-industry-digitization-index
[43] http://www3.weforum.org/docs/GITR/2012/GITR_Chapter1.11_2012.pdf
[44] http://www.strategyand.pwc.com/reports/digitization-economic-growth-job-creation
[45] http://reports.weforum.org/global-information-technology-report-2015/1-2-icts-income-inequality-and-ensuring-inclusive-growth/

Digitization Scoring: The Key Components of the Digitization Scoring, as reflected in the diagram below, includes 24 variables, grouped into 6 categories—Ubiquity, Affordability, Reliability, Speed, Usability, and Skills. While there are similarities between this and ITU's IDI, and WEF's NRI, there are also some differences.

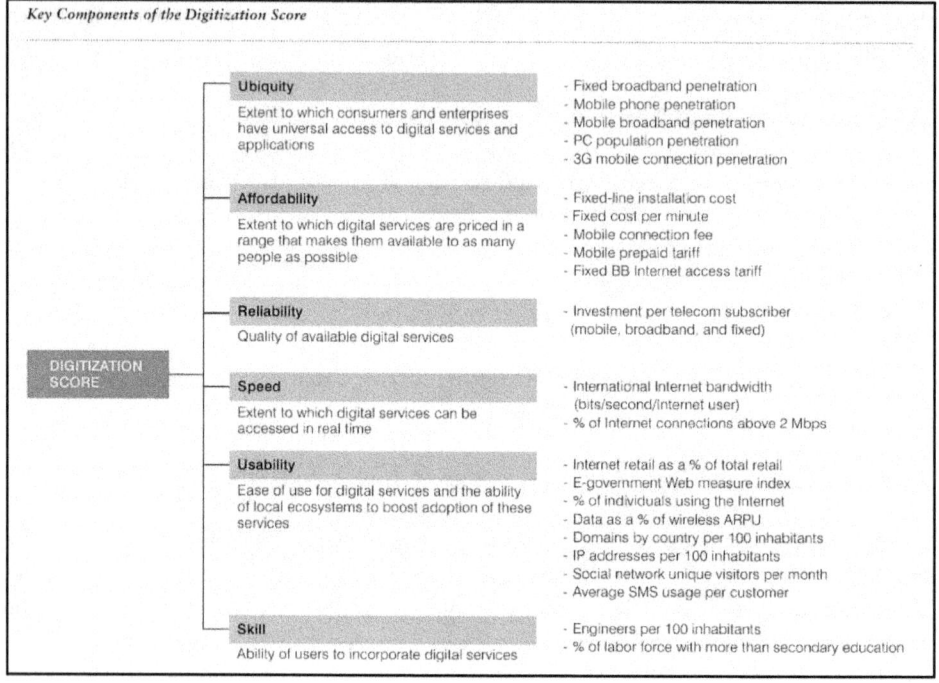

Digitization Socioeconomic Impact: The Framework for Measuring Digitization's Socioeconomic Impact is shown below, with the major components categorized as Economy, Society and Governance. This graphic also reflects the metrics for measuring the socioeconomic impact.

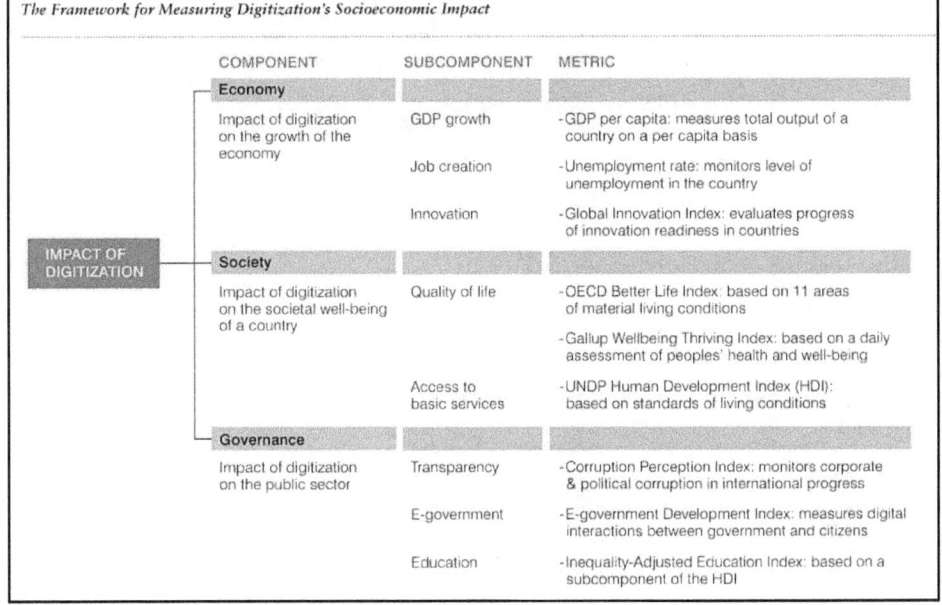

Impact of Increased Digitization: The study performed by Booz & Co in 2012, using the categories and metrics shown in the earlier Socioeconomic Impact chart, takes the analysis even further by reflecting the calculated impacts (multiplier) for each of the metrics included in the Digitization model.

The Impact of Increased Digitization

	VARIABLE	METRICS	POSITIVE IMPACT OF DIGITIZATION
ECONOMY	GDP Growth	GDP per capita: Overall	0.60%*
		GDP per capita: Constrained Stage	0.50%*
		GDP per capita: Emerging Stage	0.51%*
		GDP per capita: Transitional Stage	0.59%*
		GDP per capita: Advanced Stage	0.62%*
SOCIETY	Job Creation	Unemployment rate	−0.84%*
	Innovation	Global Innovation Index	6.27 points[†]
	Quality of Life	OECD Better Life Index	1.29 points[†]
	Access to Basic Services	UNDP HDI: Constrained & Emerging	0.13 points[†]
		UNDP HDI: Transitional & Advanced	0.06 points[†]
GOVERNANCE	Transparency	Corruption Perception Index	1.17 points[†]
	E-Government	E-Government Development Index	0.10 points[†]
	Education	Inequality-Adjusted Education Index: Constrained & Emerging	0.17 points[†]
		Inequality-Adjusted Education Index: Transitional & Advanced	0.07 points[†]

* 10 percent increase in digitization; [†] 10-point increase in digitization.
Source: Telecom Advisory Services; Booz & Company analysis

Digitization and GDP: The Booz & Co report categorizes countries into four groups: Constrained, Emerging, Transitional and Advanced. For each category, there is a varying contribution based on a 10% increase in the level of digitization. Another factor of interest is the noted GDP difference between several earlier studies that focused just on the value gained from simply expanding broadband, which is but one component of Digitization (shown on the left of the top graphic).

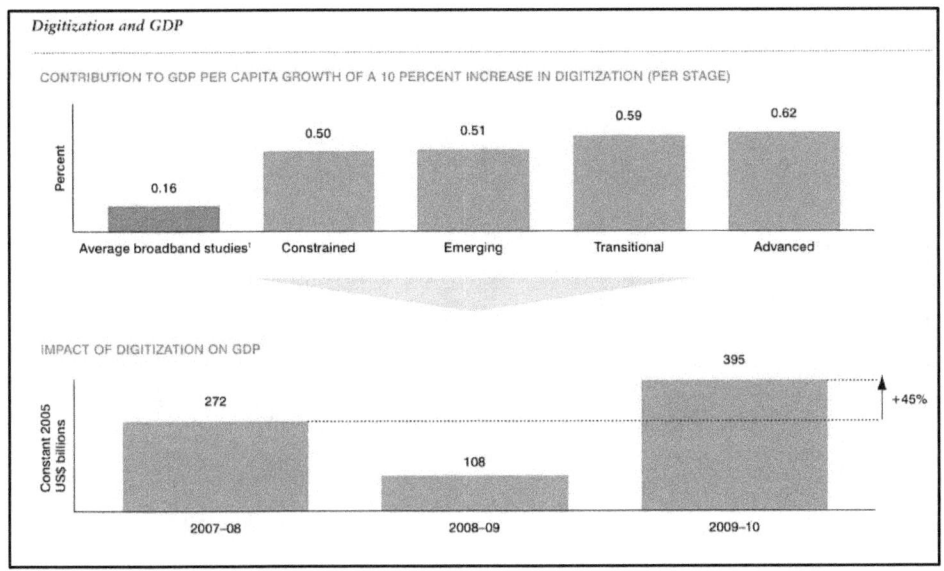

Country Ranking: Using this Digitization Index, Raul Katz and his colleagues calculated an index on the level of Digitization in 184 countries. This specific graphic relies on 2010 data.

Note specifically that those countries reflected in the Constrained category shown on the left (along with a few in the Emerging category) correlate closely with the LDCs discussed earlier in Chapter 4.

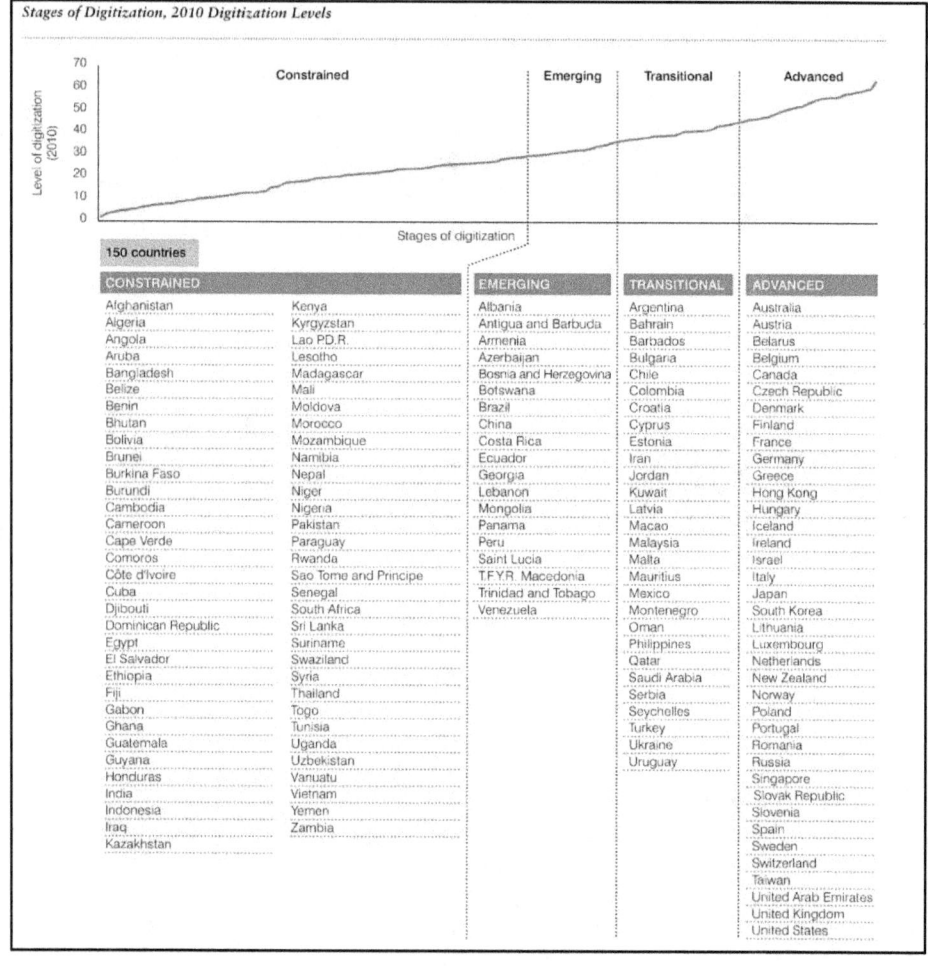

Digitization Bibliography

- **Impact of Broadband on the Economy --- 2012 --- ITU**
  - https://www.itu.int/ITU-D/treg/broadband/ITU-BB-Reports_Impact-of-Broadband-on-the-Economy.pdf
- **The two firms that are currently at the front edge of pushing the "Digitization" envelop include the following:**
  - Strategy & --- PwC – See http://www.strategyand.pwc.com
  - Telecom Advisory Services – Raul Katz. See http://www.teleadvs.com
- **Maximizing the Impact of Digitization – 2012 – Strategy&**
  - http://www.strategyand.pwc.com/media/file/Strategyand_Maximizing-the-Impact-of-Digitization.pdf
- **Using a Digitization Index to measure the Economic and Social Impact of Digital Agendas – 2013 – Raul Katz**
  - http://www.eseade.edu.ar/wp-content/uploads/2016/07/51.-callorda.pdf
- **The Varying Effects on Digitization on Economic Growth and Job Creation—A Global Perspective – 2013 – Strategy&**
  - http://www.strategyand.pwc.com/me/home/press_media/management_consulting_press_releases/details/52364755
- **Why Are 4 Billion People Without the Internet – 2016**
  - http://www.strategy-business.com/article/Why-Are-4-Billion-People-without-the-Internet?gko=cd483
- **Connecting the World – The Ten Mechanisms for Global Inclusion – 2016 – Strategy &**
  - http://www.strategyand.pwc.com/ctw
- **Connecting the World: How The Internet Can Change to Bring the World Online – 2016 – Strategy&**
  - http://www.strategyand.pwc.com/me/home/press_media/management_consulting_press_releases/details/global-internet-inclusion-ME
- **Digitization for Economic Growth and Job Creation: Regional and Industry Perspectives – 2013 – WEF/GITR/Booz&Company and Strategy&PWC**

Broadband Commission for Sustainable Development: Digitization Scorecard

In June 2017, the Broadband Commission for Sustainable Development issued a report titled, "Working Group on the Digitization Scorecard: Which Policies and Regulations can help advance Digitization[46]." While the report identified policies and regulations that help to advance digitization, these are only two elements of a broader Digitization ecosystem put forth in the following chapter. In the report's Forward by Houlin Zhau, Secretary-General of the ITU and Co-Vice Chair of the Broadband Commission, he stated the following:

"Connectivity lies at the heart of digitization. To benefit from the wealth of content and services available online, people everywhere have to have access to information and communication technologies. This is the core mission of the Broadband Commission, which has been calling for affordable and equitable access to ICTs for all people since its inception in 2010. Yet despite progress, much of the world's population still lacks access, thereby holding back progress towards the 2030 Agenda for Sustainable Development. To that end, the conclusions of this Working Group study are a valuable tool for showing how specific digital interventions may contribute to achieving the Sustainable Development Goals."

Mr. Zhau went on to state that while essential, connectivity is not enough. It must accompany responsive regulation. In addition, he put forward the importance of the broader ICT ecosystem and the essential need for cross-sector discussions to ensure collaboration and coordination. He further advocates for a multi-stakeholder approach with a focus on developing effective policy, and a legal and regulatory environment for the ICT Sector.

The ***Introduction*** of the report highlights the following:

- We are in a digital revolution which touches almost every community in the world;

- Digitalization is scaling quickly in some countries, yet more slowly in others;

- The digitization scorecard explores the digitalization readiness from a policy and regulatory perspective;

---

[46] http://www.broadbandcommission.org/Documents/publications/WG-Digitalization-Score-Card-Report2017.pdf

- The scorecard explores six countries, five sectors and foundation elements;

- Good connectivity through appropriate infrastructure is a precondition for digitization;

- The scorecard is targeted at those with an interest in digital as well as in vertical policies and regulations;

- Digitalization may contribute to achieving the UN Sustainable Development Goals.

The *Executive Summary* highlights the following:

- Most countries will benefit from a clear designation of a body responsible for digitalization through interdepartmental collaboration;

- Adequate policy frameworks should enable responsible data sharing;

- Public funding can accelerate kick-starting digitalization;

- National strategies provide clarity of vision on digitalization's critical elements;

- Education and awareness-raising are critical to effectively implement digitalization policies;

- There is no room for complacency in driving digitalization.

The report incorporates a Scorecard for six countries: Colombia, Finland, Indonesia, Kenya, Pakistan, and Singapore. While none of these countries are a Least Developed Country (LDC), Pakistan is a Least Connected Country (LCC). And while the Scorecard is limited in the breadth of countries included, where no LDC is included, there are still lessons that can be learned from Colombia, Indonesia, Kenya and Pakistan.

The graphic on the following page reflects the Summary Scorecard extracted from this report. Note specifically the differences in the "Digital Foundations" for developed countries such as Finland and Singapore when compared to Indonesia, Kenya and Pakistan. Also, note the Sector differences (what this graphic refers to as Verticals)—not only the difference between countries, but between sectors within a given country.

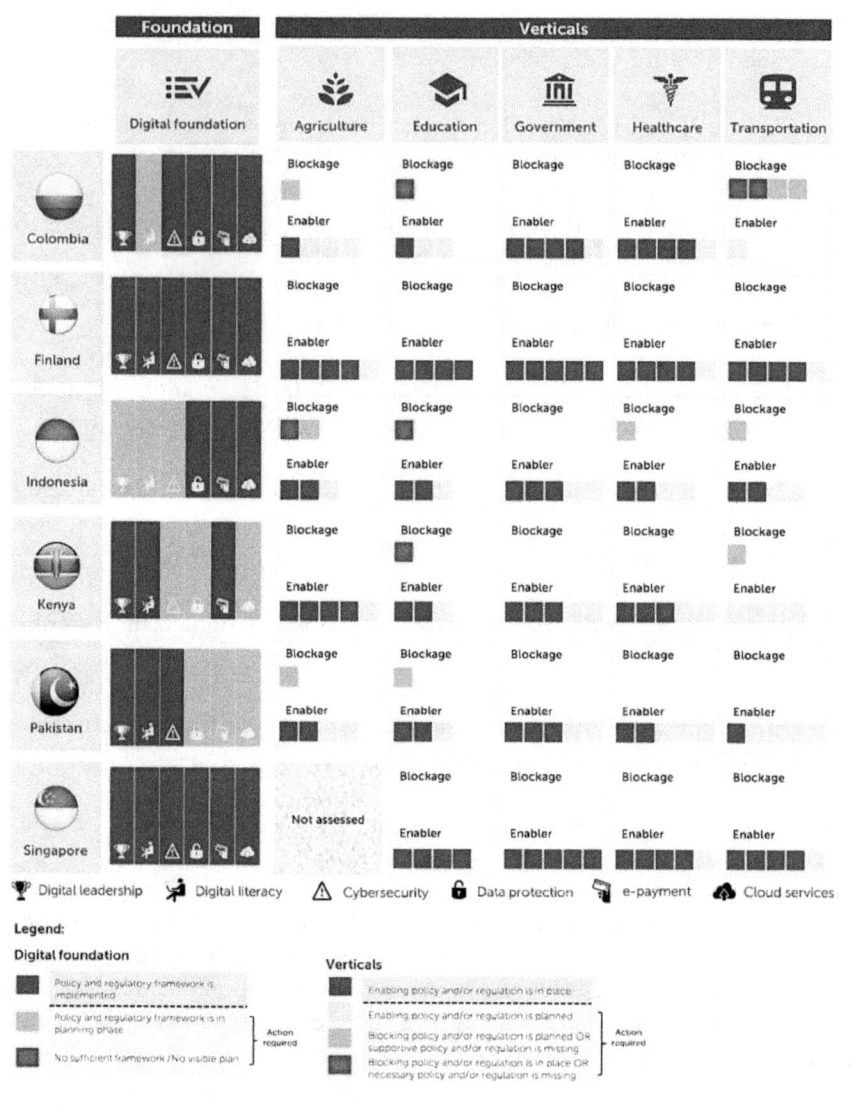

Other Ecosystem Indices

In addition to the Digitization model created and championed by Raul Katz and his colleagues, there are several other ecosystem-oriented indices that have been in place for several years, along with a couple newer additions. The following pages reflect a few of these.

Raul's Digitization relies on data from several of these indices, with his value-add being a methodology that quantifies impacts of Digitization—an essential but often-missing component.

ITU's ICT Development Index (IDI)

The ITU has tracked connectivity-related statistics for many years, and in 2008 refined this collection and reporting by creating the IDI. The IDI combines 11 indicators into one benchmark measurement. Since 2009, the ITU publishes this information annually, along with summary and analysis, in ITU's "Measuring the Information Society Report".

The Measuring the Information Society Report[47] issued in mid-2015, placed a heavy focus that built off the concluding WSIS Targets ending in 2015, and established a solid foundation for linking the IDI to ITU's new Connect 2020 Goals and Targets (discussed briefly in Chapter 3). The 2017 issue of ITU's overview is reflected in their ICT Facts and Figures 2017 publication.[48]

As stated in the report, the main objectives of the IDI are to measure the following:

- The level and evolution over time of ICT developments within countries and the experience of those countries relative to others;

- Progress in ICT development in both developed and developing countries;

- The Digital Divide, i.e., differences between countries in terms of levels of ICT development; and

- The development potential of ICTs and the extent to which countries can make use of them to enhance growth and development in the context of available capabilities and skills.

---

[47] http://www.itu.int/en/ITU-D/Statistics/Documents/publications/misr2015/MISR2015-w5.pdf
[48] http://www.itu.int/en/ITU-D/Statistics/Documents/facts/ICTFactsFigures2017.pdf

The conceptual framework for the IDI is built around a country's evolution towards becoming an information society. This is depicted using a three-stage model:

- **Stage 1: ICT Readiness** – reflecting the level of networked infrastructure and access to ICTs;

- **Stage 2: ICT Intensity** – reflecting the level of use of ICTs in the society; and

- **Stage 3: ICT Impact** – reflecting the results/outcomes of more efficient and effective ICT use.

The diagram below graphically displays this conceptual framework:

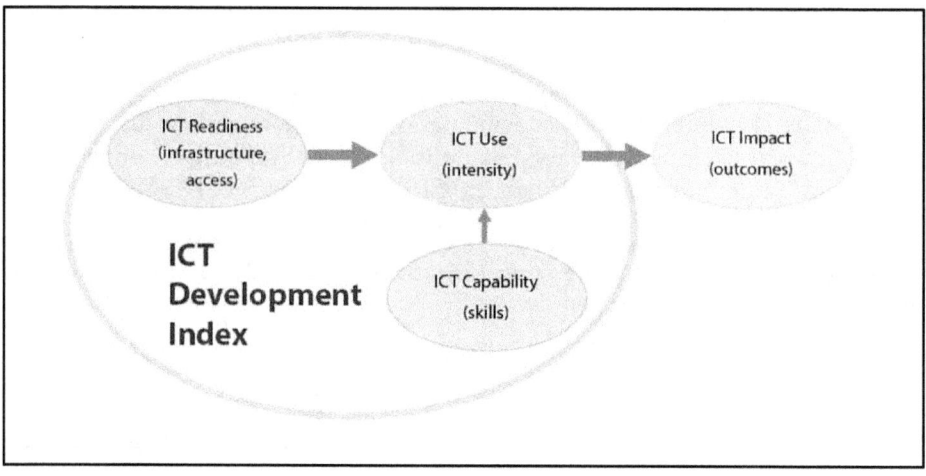

As reflected in the ITU report itself,

"The indicators used to calculate the IDI were selected on the basis of the following criteria:

- *The relevance of a particular indicator in contributing to the main objectives and conceptual framework of the IDI. For example, the selected indicators must be relevant to both developed and developing countries, and should reflect, so far as possible, the framework's three components as described above.*

- *Data availability and Quality. Data required for a large number of countries, as the IDI is a global index. There is a shortage of ICT-related data, especially usage, in the majority of developing countries. In addition, as indicators that are directly related to ICT skills are not available for most countries, it has been necessary to use proxy rather than direct indicators in the skills sub-index.*

- *The results of various statistical analyses. Principal components analysis (PCA) is used to examine the underlying nature of the data and explore whether their different dimensions are statistically well-balanced."*

The following table, extracted from the most recent ITU annual report, exhibits the indicators, reference values and weights for each of the three components in the composite IDI.

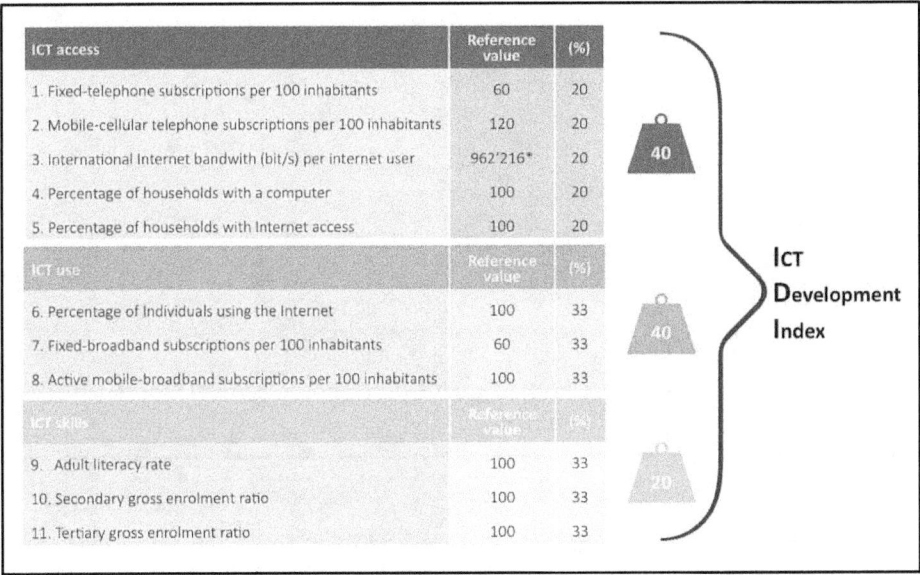

ICT access	Reference value	(%)
1. Fixed-telephone subscriptions per 100 inhabitants	60	20
2. Mobile-cellular telephone subscriptions per 100 inhabitants	120	20
3. International Internet bandwith (bit/s) per internet user	962'216*	20
4. Percentage of households with a computer	100	20
5. Percentage of households with Internet access	100	20
ICT use	**Reference value**	**(%)**
6. Percentage of Individuals using the Internet	100	33
7. Fixed-broadband subscriptions per 100 inhabitants	60	33
8. Active mobile-broadband subscriptions per 100 inhabitants	100	33
ICT skills	**Reference value**	**(%)**
9. Adult literacy rate	100	33
10. Secondary gross enrolment ratio	100	33
11. Tertiary gross enrolment ratio	100	33

The remainder of the ITU annual report includes analysis, along with country and regional information, with comparisons starting with 2010 data, as well as the most recent year the data is collected. The annual report also covers specific topics of interest at the time (i.e. the Internet of Things (IoT), and emerging dynamics with significant future implications, etc. The most recent report includes data from 2016.[49]

Of specific relevance to this book is a short discussion on the relationship between the IDI and GNI p.c. The 2015 report displays the graphic below, with text that reads as follows (extracted from pages 57-58).

> *"As noted above, one shortcoming of grouping countries by development status is that the developing countries category includes countries at varying different levels of both economic and ICT development. It is useful, therefore, also to look at the relationship IDI performance and GNI p.c."*

Chart 2.4 (reflected below), plots IDI 2015 outcomes against GNI p.c. data for 2013, shows that there is a strong and significant correlation between the two.

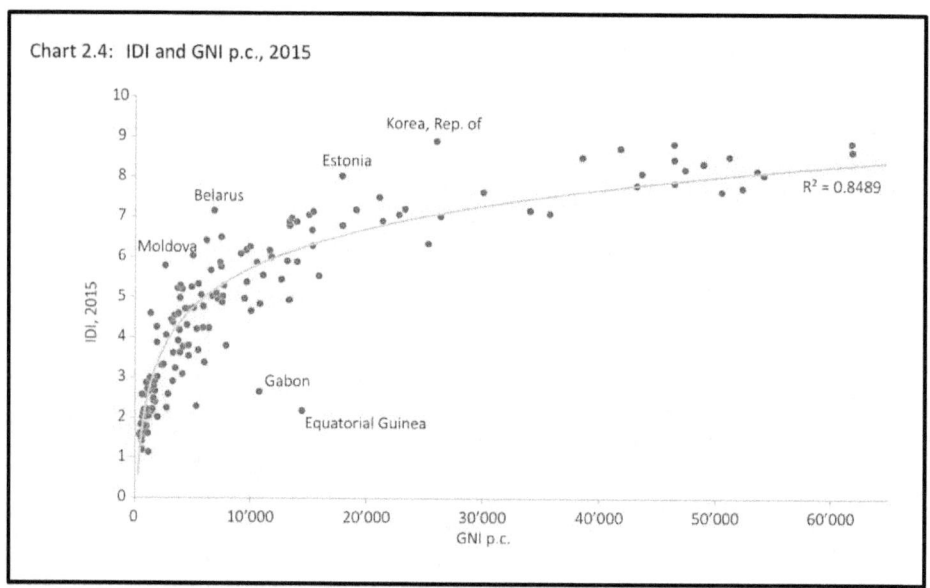

---

[49] www.itu.int/en/ITU-D/Statistics/Pages/publications/mis2016.aspx

This suggests that the level of GNI p.c. (and of disposable income within societies) influences both investment in infrastructure and the adoption of ICT services. Initiatives to stimulate ICT development may need to address the implications of this if they are to counteract the growing digital divide in those countries at the bottom of the IDI rankings. Outliers, which show significantly better or significantly weaker performance than might be expected from the data in Chart 2.4, are worth considering further. Their experience may indicate policy and investment choices, and are likely to be more or less effective in leveraging higher ICT performance. Notable outliers include the Republic of Korea, Estonia and Belarus, which outperform their GNI p.c. peers in the IDI, while two oil-exporting countries in Africa, Gabon and Equatorial Guinea, have significantly lower IDI values than their GNI p.c. peers.

In addition, as noted in the 2014 Measuring the Information Society Report, there is a strong and significant correlation between GNI p.c. and the percentage of a country's population living in urban areas (ITU, 2014b). This suggests that the concentration of population in urban areas, where costs of infrastructure investments are lower than in rural areas, may also be a significant factor influencing IDI outcomes."

WEF's Network Readiness Index (NRI)

Beginning in 2001, with a major upgrade in 2012, the World Economic Forum (WEF), in close partnership with other organizations including INSEAD, and more recently Cornell University, publishes an annual Global Information Technology Report (GITR).[50] This annual GITR presents current dynamics, with the core of the report built around the Networked Readiness Index (NRI) as reflected on the following pages. The thrust of WEF's 2016 report was "Innovating in the Digital Economy," highlighting the innovative patterns in the NRI data sets that point to policy and investment priorities with a focus on the now-emerging Fourth Industrial Revolution.

The NRI has matured over the years, with the current issue including 139 economies. It includes the state of country readiness through 53 indicators. Approximately half of these indicators are from international organizations such as the ITU, the World Bank, UNESCO and other UN Agencies. The remaining NRI indicators are from WEF's annual Executive Opinion Survey.

---

[50] http://www3.weforum.org/docs/GITR2016/WEF_GITR_Full_Report.pdf

The diagram below presents a conceptual representation of the NRI model, with the 53 NRI indicators grouped into four sub-indexes: Environment, Readiness, Usage, and Impact.

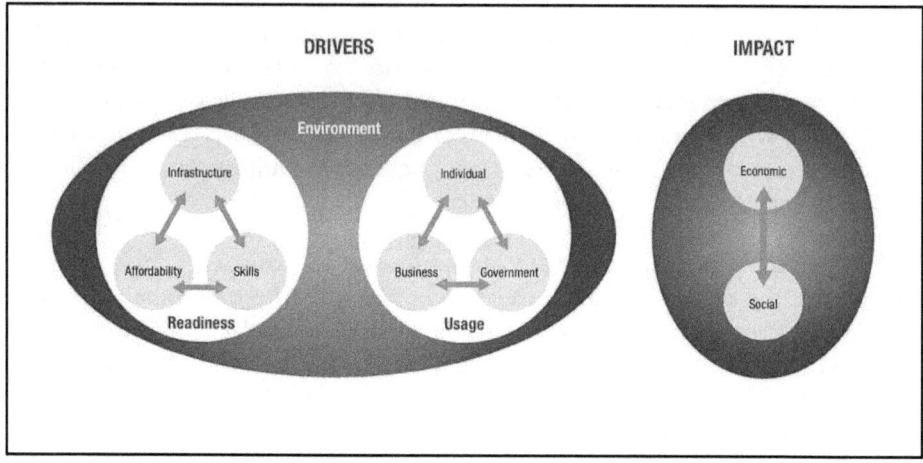

The specific alignment of the individual indicators into these four sub-indexes, along with the weight given to each, is shown on the following extract from the GITR.

NETWORKED READINESS INDEX 2016

Networked Readiness
Index = 1/4 Environment subindex
      + 1/4 Readiness subindex
      + 1/4 Usage subindex
      + 1/4 Impact subindex

ENVIRONMENT SUBINDEX

Environment subindex = 1/2 Political and regulatory environment
                     + 1/2 Business and innovation environment

1st pillar: Political and regulatory environment
1.01 Effectiveness of law-making bodies*
1.02 Laws relating to ICTs*
1.03 Judicial independence*
1.04 Efficiency of legal system in settling disputes*,5
1.05 Efficiency of legal system in challenging regulations*,5
1.06 Intellectual property protection*
1.07 Software piracy rate, % software installed
1.08 Number of procedures to enforce a contract6
1.09 Number of days to enforce a contract6

2nd pillar: Business and innovation environment
2.01 Availability of latest technologies*
2.02 Venture capital availability*
2.03 Total tax rate, % profits
2.04 Number of days to start a business7
2.05 Number of procedures to start a business7
2.06 Intensity of local competition*
2.07 Tertiary education gross enrollment rate, %
2.08 Quality of management schools*
2.09 Government procurement of advanced technology products*

READINESS SUBINDEX

Readiness subindex = 1/3 Infrastructure
                   + 1/3 Affordability
                   + 1/3 Skills

3rd pillar: Infrastructure
3.01 Electricity production, kWh/capita
3.02 Mobile network coverage, % population
3.03 International Internet bandwidth, kb/s per user
3.04 Secure Internet servers per million population

4th pillar: Affordability8
4.01 Prepaid mobile cellular tariffs, PPP $/min.
4.02 Fixed broadband Internet tariffs, PPP $/month
4.03 Internet and telephony sectors competition index, 0–2 (best)

5th pillar: Skills
5.01 Quality of education system*
5.02 Quality of math and science education*
5.03 Secondary education gross enrollment rate, %
5.04 Adult literacy rate, %

USAGE SUBINDEX

Usage subindex = 1/3 Individual usage
               + 1/3 Business usage
               + 1/3 Government usage

6th pillar: Individual usage
6.01 Mobile phone subscriptions per 100 population
6.02 Percentage of individuals using the Internet
6.03 Percentage of households with computer
6.04 Households with Internet access, %
6.05 Fixed broadband Internet subscriptions per 100 population
6.06 Mobile broadband Internet subscriptions per 100 population
6.07 Use of virtual social networks*

7th pillar: Business usage
7.01 Firm-level technology absorption*
7.02 Capacity for innovation*
7.03 PCT patent applications per million population
7.04 ICT use for business-to-business transactions*,9
7.05 Business-to-consumer Internet use*,9
7.06 Extent of staff training*

8th pillar: Government usage
8.01 Importance of ICTs to government vision*
8.02 Government Online Service Index, 0–1 (best)
8.03 Government success in ICT promotion*

IMPACT SUBINDEX

Impact subindex = 1/2 Economic impacts
                + 1/2 Social impacts

9th pillar: Economic impacts
9.01 Impact of ICTs on business models*
9.02 ICT PCT patent applications per million population
9.03 Impact of ICTs on organizational models*
9.04 Knowledge intensive jobs, % workforce

10th pillar: Social impacts
10.01 Impact of ICTs on access to basic services*
10.02 Internet access in schools*
10.03 ICT use and government efficiency*
10.04 E-Participation Index, 0–1 (best)

The NRI portion of WEF's annual report ranks each of the countries by 10 sub-indexes on the above table, and compares the country profile to the income group within which the country fits. The four levels include: Low-Income Group, Lower-Middle Income Group, Upper-Middle Income Group, and High-Income Group.

The following graphic is a WEF produced Radar-Chart extracted from their annual GITR—here the diagram reflects the profile for: 1) the average of the Emerging and Developing Economies, and 2) the average of the Advanced Economies. I have annotated the average profile for Low-Income Countries.

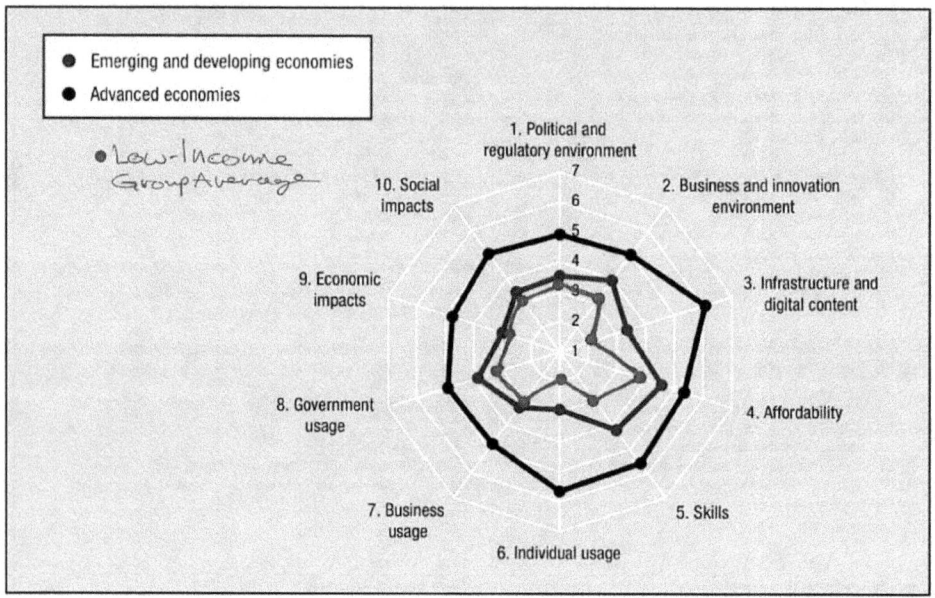

Two key values of this data are:

1) Its historical perspective that captures changes from year to year, identifying where progress has been made, and;

2) Comparison to other countries in the same economic category—identifying possible opportunities to leap forward by copying successes of others.

But perhaps even more important is the potential analytical value to quickly assess a specific country's rating, to drill down to identify target areas needing attention in a logical sequence, and then to track impacts of these changes, should an agenda for improvement be undertaken.

A4AI's Affordability Drivers Index (ADI)

The Alliance for Affordable Internet (A4AI) published its latest 2017 Affordability Report in February 2017.[51] This latest issue is an update from reports issued in 2014 and again in 2016. The report keys off the Sustainable Development Goals, specifically SDG target 9c, that calls for universal and affordable access in the world's least developed countries by 2020. The report explored the affordability environments of 57 countries, taking into consideration the effects of poverty and income inequality, as well as gender inequality relative to access. The 2017 report looked specifically at how poverty, income, and gender inequality are masking the true state of affordability. It put forward several recommendations to address these inequalities.

A4AI Recommendations (from Executive Summary)
- Employ Public Access Solutions to Close the Digital Divide
- Foster Market Competition Through Smart Policy
- Implement Innovative Uses of Spectrum through Transparent Policy
- Take Urgent Action to Promote Infrastructure and Resource Sharing
- Make Effective Use of Universal Service and Access Funds
- Ensure Effective Broadband Planning Turns into Effective Implementation

The ADI looks at policies, incentives and infrastructure investments in place across these 57 countries (developing and emerging), to assess the nature of their implementation.

The ADI data is broken into two Sub-Indexes—one for Access and one for Affordability. There are 29 Access-related Indicators, along with Affordability Indicators from the World Development Indicators Database. The World Development Indicators Database is also relied upon for calculating Affordability.

The A4AI has adopted an affordability target of 5% of an individual's income for Internet access. This is consistent with that of the Broadband Commission for Sustainable Development's target. The report found that of all 57 countries included in their analysis, none with populations at or below the poverty level of US$ 3.10/day met this target. The total population of those living in poverty in these 57 countries is placed at 1.9 billion. This highlights yet again the challenge of the 48 LDCs, where an average of 70% live in extreme poverty—less than US$ 1.90/day.

---

[51] http://1e8q3q16vyc81g8l3h3md6q5f5e.wpengine.netdna-cdn.com/wp-content/uploads/2017/02/A4AI-2017-Affordability-Report.pdf

GSMA's Connected Society: Mobile Connectivity Index (MCI)

In July 2016, the GSMA published a report, along with an interactive website, that measures performance of 134 countries. This first Mobile Connectivity Index (MCI) is built on 2014 data[52]. The 134 countries account for over 95% of the world's population.

As stated in the Executive Summary of their 2016 Report...

> *"This new product has been developed with a clear and simple motivation – to support the efforts of the mobile industry and the wider international community to deliver on the ambition of universal access to the Internet."*

The MCI was designed...

> *"...as a tool to help focus the efforts and resources of the mobile industry and wider international community on the right projects in the right markets at the right time, so progress towards universal access can be as swift and economically sustainable as possible."*

The index measures how the key enabling factors for mobile connectivity differ across markets. It seeks to answer two key questions:

- What factors need to be in place to create the right conditions for supply and demand to flourish?

- How can countries at the beginning of the journey towards universal service access benefit from the experience of those further along?

The report ties directly to supporting the SDGs, and identifies four key enablers driving adoption of the mobile Internet.

---

[52] https://www.gsma.com/mobilefordevelopment/programmes/connected-society

> **Key Enablers**
>
> - **Infrastructure**—the availability of high-performance mobile Internet network coverage
> - **Affordability**—the availability of mobile services and devices at price points that reflect the level of income across a national population
> - **Consumer Readiness**—citizens with the awareness and skills needed to value and use the Internet and a cultural environment that promotes gender equality
> - **Content**—the availability of online content and services that are accessible and relevant to the local population

Each of these four enablers is given a weight factor of 25% in their Index, with 38 specific indicators included in the MCI.

The GSMA report goes on to highlight the linkage between broadband availability and economic impacts by citing several studies—a topic covered in more detail in the earlier discussion on Digitization that relied upon the analysis by Raul Katz and his colleagues.

The GSMA Mobile Connectivity Index[53] was updated in 2017, and is supported by an online interactive capability where one can drill down on a specific country and get basic information.

---

[53] http://www.mobileconnectivityindex.com/#

Value Derived from the Ecosystem Data Sets

The above-referenced sources periodically collect and publish new data sets that collectively serve as valuable assessment tools. While each has somewhat differing sets of data and adopts different approaches in grouping the core data sets, there is also some overlap from key sources such as ITU.

These Ecosystem data sets are invaluable sources for quick diagnostics of a country's on-the-ground status. They provide preliminary data sets for identifying and assessing topics needing further drill down. Further, they are invaluable in examining differences and similarities between countries, along with changes over time.

The work undertaken by Raul Katz and his colleagues is perhaps the most exciting and promising for the future. Their work started down the path of establishing sector-specific value-add from Digitization.

This is an area of research that needs to be expanded and taken further—with the ultimate goal to add focus on identifying and refining more prescriptive approaches leading to targeted action.

...experiences and observations from the author

Within two weeks after retiring from USAID, I had my first consulting engagement. The project was a three-week Internet for Economic Development (IED) Assessment in Morocco. After Morocco, an ICT Assessment for Sri Lanka followed. For both I was assigned to a much larger team, with my primary focus being connectivity.

The projects were to work with U.S. Embassies and USAID's Missions, to assess the local country landscape, and explore opportunities for including ICT within their country-development portfolios. I did my part for both countries and turned in my materials, which were then consolidated into full contractor reports.

In early 2000, the contractor asked me to return to Morocco to give an informal evening presentation about our earlier assessment at a planned U.S. Ambassador-hosted event at his residence. Over 100 people we had met with during our Assessment were to attend.

I read the completed final report for the first time on the plane going over to Morocco. I got sicker with each page I read. It was a collection of all our input—with virtually no structure, theme, or compelling message. As the flight proceeded towards Morocco, I was forced to create these for my presentation.

A friend living in Morocco who was on the Assessment team, Karl Stanzick, picked me up at the airport... greeting me at the airport with a message, "The U.S. Ambassador wants to see you as soon as we can get there." I recall walking into the Embassy, meeting with the Ambassador, who immediately started to talk about the report and the next evening's event. He too had tried reading the report and was wondering what it said. We shared our mutual dismay. I then proceeded to provide him with a structure and message that I had reconstructed from the report while on the plane. He was relieved. And it carried our discussion the following evening.

Upon returning to Washington DC, I got in touch with my contractor and USAID with a message—we needed to get our act together and develop a structured process and framework for future engagements. We did this, and began using this model in our subsequent country assessments with positive results. We continued to refine the process and structured framework.

On reflection, it is interesting to recall that for the most part in 2000, our work was in new territory. There was ITU data, and some other reports as well—but there were very few and those were basic. Most data sources referenced in this chapter simply didn't exist. We were largely on our own to figure out the in-country dynamics and identify areas where USAID could contribute.

The Development through Digitization (DtD) model put forth in the following chapter represents my most recent structure after being engaged at some level in 30+ countries. I have no doubt this proposed DtD model can be improved upon, and hope others working in this space will do so.

CHAPTER 6

– The Second Bookend – Development through Digitization (DtD) Model

This chapter represents the second bookend. The core rationale for writing this book is to pull together my observations, experiences and lessons learned. And to this, add research that puts forward an ecosystem-oriented construct to move the discussion and practice forward. This chapter puts forward a Development through Digitization (DtD) model for achieving on-the-ground results in developing countries, with an emphasis on the LDCs.

The key consideration is seeking a more systemic approach capable of driving socioeconomic development, hence the title, **Development** (as the primary focus) "through" **Digitization** (as a comprehensive ecosystem-approach for achieving development impact).

The chapter is divided into two Sections. First is a short overview of the six components of the DtD model, followed by a more in-depth section that delves deeper into each of the elements within the six DtD components.

DtD Model: Overview

One of the key differences in developing countries is that in the mid-upper and middle-income categories, there is most often a sufficient level of demand for Internet access already in place, though perhaps not fully satisfied—certainly not optimized relative to maximizing socioeconomic impact. When access is provided, the demand is enough to justify the build-out investments and support ongoing operating costs. However, here too, there must be a focus on expanding local demand via building locally relevant content and services. In the LDCs, typically there is little demand as a starting point. Consequently, attention must focus on putting into place targeted local value-added content and services that serve to kickstart build-out and user adoption.

The Digitization model is structured into six complementing components as shown in the diagram on the right. Within each of these six components there are several elements. Attention to each of these elements is necessary to achieve the sought-after socioeconomic value.

The most direct linkage of Digitization relative to achieving socioeconomic impact was put forth in the last Chapter covering Raul Katz's Digitization Index. His work classifies countries into four categories—Constrained, Emerging, Transitional, and Advanced.[54] His "constrained" country category corresponds closely to the LDCs.

The next few pages provide a quick overview of each component reflected in this DtD model, under a paragraph labeled, "Focus". In addition, for each of these six components, there is a bulleted list of "Key Elements." The second section of this chapter, "DtD Model: Drill Down" provides more detailed discussion for each component and the supporting elements.

DtD Component:
Nexus for Change

Focus: A national level commitment is required to pull together government, international donors, and international and local private sector firms. The target result is establishing a shared foundation upon which to support development of a comprehensive Digitization agenda—one that becomes a national priority, and where a comprehensive Digitization program serves as a valuable contributor for socioeconomic change.

Nexus for Change: Key Elements:

- National government commitment

- In-country advocate-center for change

- International, regional, and national coordination and collaboration

---

[54] Using a digitization index to measure the economic and social impact of digital agenda (http://www.eseade.edu.ar/wp-content/uploads/2016/07/51.-callorda.pdf)

- High level ICT Steering Committee comprised of both public and private sector representatives

- Donor coordination on all things ICT

- Integration of public and private sector initiatives

- Key reference point for advocacy and pulling together national Digitization vision, plans, monitoring status

DtD Component:
Shared Vision and Commitment

Focus: A shared public and private sector vision, priorities, and commitment is essential for leveraging all-things-digital to support socioeconomic growth. This collective vision and commitment are captured in a series of products that provide the base line for shaping the future. These must be defined in adequate detail to support monitoring and evaluation for periodic refresh and refinement.

Shared Vision and Commitment: Key Elements:

- National Digitization Vision

- National Digitization Strategic Plan

- National Digitization Tactical Plan

- M&E plan with periodic reviews

- Commitment on liberalizing the telecom market

- Public and private sector buy-in

- Periodic refresh of the Vision an all Plans

- Collaboration with sub-regional and regional ICT-related organizations

DtD Component:
Enabling and Facilitating Environment

Focus: A comprehensive legal and regulatory environment that is transparent, predictable, and well-staffed in numbers and capabilities will create an enabling environment. This facilitates private sector infrastructure investments, especially on expanding the rollout and adoption of affordable Internet to reach rural populations.

Enabling and Facilitating Environment: Key Elements:

- New telecommunications law as needed

- Removing legal and regulatory constraints and regulatory strengthening with full independence from outside influence

- Updated and effective spectrum management to better leverage new emerging wireless technologies

- Lowering ICT-related taxes to help expand affordability

- New-updated and active universal service fund (USF)

DtD Component:
Ubiquitous and Affordable Access

Focus: Ubiquitous and Affordable Access is another fundamental requirement. This requires access to shared international broadband, establishment of a shared national fiber backbone, MNO and ISP build-out, and distribution for delivering broadband at locally affordable rates.

Ubiquitous and Affordable Access: Key Elements:

- Shared carrier access to international and national broadband backbone

- Open and competitive telecommunications market

- Adoption of low-cost, low-power connectivity networks

- Support for new and innovative business models for expanding rural broadband

- National Internet Exchange Points (IXPs) to minimize international transit of local content

- Sustainable rural broadband build-out

- Rural community access centers

- User-device purchasing programs (desktops, laptops, tablets and smart phones.

DtD Component: ICT-Related Knowledge and Skills

Focus: Commitment and active engagement in building local ICT knowledge and skill sets sufficient to launch/expand a local ICT industry is needed. This industry supports the public, private, and individual ICT adoption and future growth.

ICT-Related Knowledge and Skills: Key Elements:

- ICT job skills development

- Directly engage international firms such as Cisco, Microsoft, Intel, IBM, HP, Google, Facebook, etc.

- Technology training and certification programs

- ICT-related soft skills training

- Innovation and entrepreneurial programs and laboratories

- Local university ICT-related certification and degree programs

DtD Component: Relevant Content and Adoption

Focus: Development and expansion of locally-relevant/locally-hosted, value-added content and services is another necessary element. It strategically targets socioeconomic advancement, along with individual, government, and business adoption of ICTs across the public and private sectors. This component is the essential ingredient for achieving the sector-specific value-add delivered through digitization.

Relevant Content and Adoption: Key Elements:

- Adoption of ICT by local governments and businesses

- Support for local ICT sector and ICT-enhanced services imbedded into local sectors

- ICT value-add focused on supporting and achieving Sustainable Development Goals (SDGs)

- Expand use of applications/content for social and economic value

- Develop applications and content for advancing public-sector services: government, education, health, and agriculture, and private sector services such as agriculture, financial services, trade, etc.

- Promote mobile applications and local mobile marketplace

DtD Model: Drill Down

The remainder of this chapter drills down into each of the six DtD Components, including short explanations of the key Elements within each of these Components.

DtD Component: Nexus for Change

Having Digitization, or even expanding affordable Internet access, as a national priority is not yet commonplace within the LDCs. And even where it is, moving it from a stated ideal to reality requires a high level of across-the-country commitment. This commitment is often lacking due to either limited vision, or in some cases, due to local vested private interests of those in a position that may be threatened by an open and transparent Digitization approach. Naturally, other national priorities are also factors in diverting attention away from pursuing Digitization.

> **Nexus for Change:**
> This is the essential core requirement—the foundation stone for positioning Digitization as a critical component for supporting socioeconomic advancement.

Commitment needs to come from the Office of the President or Prime Minister, with full support of Parliament, and the various government Ministers.

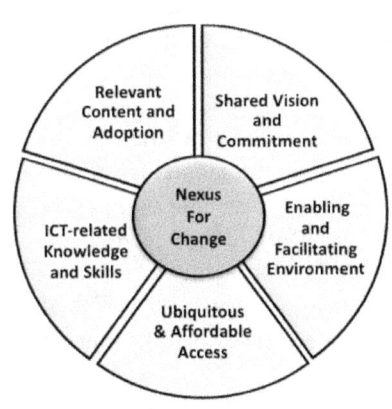

Implementing mechanisms are required, but often a key restraint is the lack of a strong commitment from the highest levels of government, along with support from the major high-tech private sector.

A place to start is establishing a short-term initiative to assess the current in-country situation, and to compare the local situation to other countries in the region, including those that have been successful. The ecosystems discussed in Chapter 5 hold value for undertaking this endeavor.

This initial assessment forms the foundation for creating an agenda that establishes not only the country-level Nexus for Change, but also for shaping a roadmap to bring about the needed changes. It should include establishing targeted results with timeframes and milestones for achieving results.

If this first essential component is not put into motion, the remainder of the contents put forward in this Chapter will likely produce limited value—perhaps even no value, with regards to leveraging Digitization for advancing socioeconomic development.

Elaboration of the Nexus for Change Key Elements:

- **National government commitment**—For developing countries, and especially LDCs, there are often significant numbers of high priority topics needing attention. Frequently, these are viewed as more immediate in nature than Digitization. However, if the country is to move forward and leverage the socioeconomic power of Digitization, there needs to be a commitment at the highest levels of government. The results may not be immediate, but getting started is essential. Ideally this commitment will be at the Presidential and Legislative levels. The commitment doesn't need to have extensive details, but rather the key is that Digitization be recognized as a national priority, with a delegated entity having lead responsibility for pulling together a team and establishing an agenda.

- **In-country Center for Change**—The first order of business after formalizing a national commitment, is to establish a core entity that serves as the national advocate. It need not be a large staff, but it must go beyond symbolism, and be staffed with qualified people capable of creating an agenda and moving forward. Location is important. This could be located within the President's office, a national development agency, or an economic development ministry. This Center for Change initially will want to focus on building inside support and seeking support from the outside—possibly from the donor community.

- **International, regional, and coordination and collaboration**—A key focal point for the Center for Change is to serve as the single, high level point of contact and coordination for all things relating to Digitization. This should include serving as a single point-of-reference for the international and regional donors, as well as serving as a single point for national coordination and collaboration. This role is not a "control" type of engagement, but rather to build and move the agenda towards national prioritization, donor coordination, and moving towards establishing national strategic and tactical plans. A key element here is to have the Center for Change lead the donor coordination around a national agenda.

- **High level Digitization Steering Committee comprised of both public and private sector representatives**—While the Center for Change will likely consist of a small staff, there is the need for the Center to reach out and establish support from, and collaboration with, both the public and key private sector. A logical construct for this is to establish an executive-level steering committee of key executives from the government, selected high-tech firms, and potentially contributing universities. Consideration should be given to having the private sector executives serve on a rotational basis.

- **Coordination-Integration of public and private sector initiatives**—In launching the new Center for Change, an early focal point should be to build an inventory of key public and private sector initiatives that are already underway. The Steering Committee members will be key in bringing in the needed participants and building this inventory. This should result in a core document that catalogs and provides a narrative description of ICT-related initiatives, systems, education, etc., which are currently in place or being planned. This is the starting point.

- **Key reference point for advocacy and pulling together a national Digitization vision, plans, monitoring status**—Another essential role of the Center for Change is to take this core set of initiatives and begin building a consensus-based vision, and subsequent plans. This is put forward as the next two Components—Shared Vision and Enabling Environments.

...experiences and observations from the author

Using **Armenia** as an example of my experiences with this component, there were significant successes in establishing both the Nexus for Change and the subsequent Shared Vision and Commitment. This dynamic took place in Armenia during the 2000-2002 timeframe. Following are key elements of this success:

National Steering Committee—to support the process, a steering committee was formed to include representation from the Office of the President, Parliament, government Ministries, and key private sector high tech firms operating in Armenia.

ITDSC—to support the day-to-day commitment, an Information Technology Decision Support Committee (ITDSC) was established to support the plan and the steering committee by monitoring progress, identifying and assessing key topics, putting decision documents in front of the committee for consideration, etc. The ITDSC was located within the Armenian Development Agency (ADA) and supported by the local USAID Mission through the Program Technology Transfer (PTT) program.

One of the first orders of business was an agreement between USAID, the GoA, and the World Bank, to build Armenia's first National ICT Strategic Plan. This prioritized the near-term action items and helped ensure there was coordination and buy-in amongst the international donor community engaged in Armenia.

...experiences and observations from the author

Unfortunately, not every engagement in developing countries yields positive results. Where national-level government support is missing, success in the other components reflected in this DtD model become impossible to pursue. Some of this is caused by timing, lack of vision or overriding local priorities.

In **Eritrea,** we undertook several opportunities in the early 2000s to advance their leveraging of ICTs. There was a specific focus on skills development through the University of Asmara. This included setting up two Cisco Academies, two computer labs—one being Microsoft and the other Open Source, plus a digital library.

But when it came to liberalizing the telecom market, our efforts had limited success. We were able to provide support for corporatizing EriTel, the monopoly provider, such that it could retain its earnings to support build-out. But we were unable to gain approval of the government to move towards privatization and introducing competition into the marketplace. Expanding international Internet access was limited to satellite-based access put into place by USAID's Leland Initiative, as we could not get local support for establishing a landing station capable of connecting to an existing undersea fiber passing by Eritrea. Recent ITU data reflects Eritrea's level of connectivity remains at the bottom of virtually all countries.

In **South Sudan,** I had the opportunity to engage within a few months of South Sudan gaining independence from Sudan. There was strong support from across the international development community, and even the private sector, to expand telecommunications. But the lack of in-country support from their national leadership to chart a future direction limited most all future engagements. The continued political unrest is another factor that takes a higher priority and limits pursuing anything in the Digitization area. South Sudan, like Eritrea, remains at the bottom of the ITU scale regarding connectivity.

DtD Component:
Shared Vision and Commitment

This component builds off the country-level commitment for change. Here the focus shifts towards creating more refined strategic and tactical action plans for achieving translating the commitment into action.

To accomplish the essential elements of this component, the earlier commitment must be translated into plans, with local and international resources secured to support action.

> **Shared Vision and Commitment:**
>
> To gain advantages of Digitization, it is essential that all the ecosystem elements be architected into an integrated whole and actively supported by both the public and private sectors.

Ideally this can be achieved through a high-level team having public sector political support, along with support from the private sector, and ideally with external support from the international donor community.

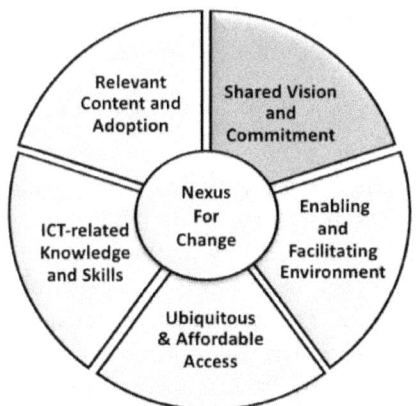

There are several priority topics here, but the core of this component is built around constructing a National Digitization Strategic and Tactical Plan. This includes not simply the infrastructure build-out, but also building local skills, along with local content and services. With regards to content, the initial focus logically includes placing a priority on providing a range of government-provided services. These services include education, E-government, health, citizen services, business-support, etc.

It is essential that this planning include a sufficient level of detail to support funding and implementing decisions, as well as tracking the progress. If the country has a Universal Service-Access Fund, the national plan provides the guidance and priorities for disbursing these funds.

As reflected under the Nexus for Change, it is also crucial that a formal organization with public and private sector participation be put into place, and that this organization is staffed with the core competencies and skill sets to carry out each of the following elements.

Elaboration of the Shared Vision and Commitment Key Elements:

- **National Digitization Vision**—An essential place to begin is expanding affordable Internet. But the Digitization vision must go well beyond connectivity and include a focus on gaining value out of all the components reflected in this Digitization model. Here the priority needs to articulate a country-specific vision and prioritization across the entire Digitization ecosystem that guides both the direction and the target results. The question of, "What does the local Digitization vision look like 5-10 years in the future?" needs to be answered.

- **National Digitization Strategic Plan**—Here there is a shift to a high-level strategic plan for achieving the vision. The Digitization model presented in this chapter has value in identifying topics to address at the strategic level, as well as at the more detailed tactical level (next bullet). The strategic plan focuses on gaining broad agreement on the "what" is included in the agenda, with the following tactical planning focusing on "how" this agenda is to be achieved. A valuable starting point is to launch the planning process via a Strength Weakness Opportunity Threat (SWOT) analysis. Effectively the Digitization Plans (strategic and tactical) reflect an upgrade to any existing ICT or Broadband Plans, should they already exist. But equally important is that they do not become simply documents, but rather a collective process with buy-in for developing the documents that lead to action.

- **National Digitization Tactical Plan**—The Tactical Plan provides a level of detail that defines project plans, schedules, resources needed to achieve the target results, funding, human resources, and a timeframe. Further, the level of detail needs to be such that progress can be monitored along with periodic independent evaluations. Consideration should be given to using outside consulting firms familiar with facilitating this specific topical area—at least for the first planning effort. This is an area where the international donor community can provide subject matter expertise to launch and build local capacity for moving this forward. But it is also essential the local Steering Committee ultimately own both the Strategic and Tactical plans, and that the process builds in-house capabilities for developing future iterations.

- **M&E plan with periodic reviews**—This element consists of putting in place a feedback loop to ensure monitoring and reporting of the progress made against the plan. The evaluation should be independent and initially done on the order of every 12 months. Once the Steering Committee overseeing the overall Digitization effort reaches a comfort level with the plan's direction and progress, the M&E can occur every 18-24 months.

- **Periodic refresh all Plans**—The tactical plan should be on the order of a 5-year time horizon. The M&E process should serve as the driver for making the needed adjustments in refreshing the plan. Here again, an outside consulting firm can assist in this process the first time, with local resources conducting the evaluation after the initial update.

- **Focus and commitment on liberalizing the telecom market**—Where the next component focuses on creating an enabling and facilitating environment, here in the shared vision and commitment component, a commitment to market liberalization is needed. This should include the entire value-chain (e.g., distribution, national backbone, and international gateway).

- **Public and private sector buy-in**—While the vision, strategic, and tactical plans are essential documents, it is equally important that the process and agreements on the final content of these products be as participatory as possible. This buy-in is every bit as essential as the details reflected within the planning documents. Documents are just that, documents. But for action, there must be active buy-in with support that results in the commitment of resources and subsequent implementation.

- **Close collaboration with sub-regional and regional ICT-related organizations**—The dominant focus for establishing a shared vision and commitment is at the country level. However, there is also a value-add in pursuing collaboration and integration with neighboring countries. This is especially critical in the connectivity arena, where there are likely sub-regional or regional broadband backbone initiatives that traverse multiple countries from a nearby undersea fiber landing. Digital financial services is another area where supporting mobile cross-border transactions holds the potential for expanded market access. Neighboring countries may also be more advanced, and there may be the opportunity to learn and gain support from their experiences and successes, thereby avoiding pitfalls.

...experiences and observations from the author

Building off my earlier **Armenia** discussion on Nexus for Change, the next focus in the Digitization model is to leverage the mechanisms put into place—to set the direction and ensure it is kept current. This includes:

National ICT Strategic Plan—early in our USAID engagement in Armenia, there was collaboration and support for the development of a National ICT Strategic Plan. The World Bank assisted us with support from the Office of the President.

M&E and Refresh—after the strategic plan was in place for 18 months, an independent review was undertaken by a local NGO that examined progress and non-progress, as well as recommended changes for consideration in a refresh of the plan. Armenia's Information Technology Decision Support Committee (ITDSC) managed this process with support from USAID. The result was an updated and refined National ICT Strategic Plan approved by the Steering Committee.

Country-Led Donor and Project Coordination—One of the significant values in country-led vision, strategic and tactical plans, and projects, is that they facilitate donor coordination. Country leadership goes well beyond a peer-to-peer coordination and collaboration between donors. The donor coordination must include project-level coordination to ensure consistency and integration of the various parallel initiatives, tool set consistency, data naming conventions, etc. The ITDSC was active to ensure there was continuity and coordination across the development community on all-things-digital undertaken within Armenia. This added country ownership and priority, and ensured integration of parallel projects. The initiatives were "pulled" into place and owned by the country and not "pushed" into the country by the donors.

...experiences and observations from the author

Another example of shared vision and commitment:

Indonesia—USAID's engagement through the GBI Program, focused on building a broadly accepted Indonesia Broadband Plan (IBP). The plan was signed in the fall of 2015, within the first week of the then newly-elected President being sworn into office. The IBP serves as the blueprint for expanding their broadband infrastructure through a reinvigorated USO Fund, with nearly US$ 500 million disbursed within the first year. It also incorporated broad support for delivering expanded government content and services over the Internet through the various Ministries.

DtD Component:
Enabling and Facilitating Environments

Most often the LDCs have affordable voice and text access that reaches into rural areas through mobile networks. However, there are often constraints associated with expanding rural Internet. The primary constraint is a financial imbalance between the costs of delivery (CapEx and OpEx) relative to an initial limited demand. This is caused by: 1) the higher cost of delivering Internet access due largely to longer distances of the backbone; and 2) the lower level of rural-demand from the lack of written language and digital literacy, and locally relevant content.

> **Enabling and Facilitating Environment:**
>
> For most LDCs, "enabling" an environment for Internet expansion doesn't go far enough. Rather, there is the need for a more aggressive, "facilitating" approach aimed at making changes happen to both the demand and supply side of the equation.

One toolset to fill this cost-revenue gap is a Universal Service Fund (USF). However, USFs have had a history of not being effective. This is typically caused by not having a national plan in place that provides an agreed-upon national architecture for identifying and prioritizing the distribution of these funds. Two key areas needing attention include: 1) lowering the cost of delivery via shared broadband infrastructure investments, along with lowering of consumer-related taxes; and 2) enhancing the richness of relevant content such that it drives the demand for Internet access (Component 6).

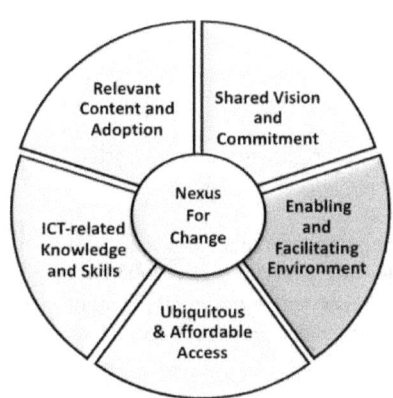

For LDCs, the focus for this component must move beyond creating an environment that is simply "enabling." Rather, a more aggressive approach is needed—one that creates a "facilitating" environment that incentivizes and encourages Internet expansion. It requires making things happen on both the demand and supply side. It is promoting the development of public and private sector content, and enhancing the language and computer literacies of local and rural populations.

Elaboration of the Enabling and Facilitating Environment Key Elements:

- **New telecommunications law, as needed**—Historically, in many countries the telecommunications sector was initially owned and operated by the national government as part of their Post, Transport and Telecommunications (PTT). Most governments have since moved to privatize the telecommunications sector through a combination of legislation and auctions. This was especially the case for mobile operators, which dominated the industry for the last 15 years. There is likely the need to review the current laws to ensure optimal competition and to update the foundation for managing the future direction of the industry. A key component in any upgrade is to establish delegations to the regulatory authority such that there is minimal need for future legislative engagement to address issues needing attention.

- **Removing legal and regulatory constraints and regulatory strengthening**—Typically the telecommunications law is the foundation upon which the regulatory details add to the refinement that shapes the industry. A key component is to establish consistent legal and regulatory authorities with implementing regulations. This provides predictability and transparency—allowing the industry to pursue expansion with minimal ambiguity and intervention by the regulator. Another key focus is to develop in-country regulatory capacity sufficient to address ambiguities on case-by-case situations. It is also essential to structure the regulator as an independent entity with minimal political influence.

...experiences and observations from the author

While the USFs are one approach for expanding rural access, another approach is to incorporate a rural build-out requirement into the license of a monopoly provider.

This was the case in **Armenia** with ArmenTel, the monopoly mobile provider. USAID supported a study by an international independent legal firm to explore whether or not ArmenTel was in compliance with their licensing agreement on rural build-out. They were not. The Minister of Justice subsequently took ArmenTel to court and ArmenTel was found to be non-compliant.

The agreed-upon court settlement allowed for the licensing of a second mobile operator, which for the first time, introduced competition into Armenia in the telecommunications sector. The result was an expanded mobile coverage and uptake, along with reduced pricing.

- **Updated and effective spectrum management to better leverage emerging wireless technologies**—One of the rich opportunities of the last decade has been the technology migration from analog to digital transmission. However, with this opportunity comes a list of challenges, from migrating technologies, reallocation of spectrum licensing, realigning freed spectrum, etc. The target benefit in migrating from analog-to-digital transmission is that there is typically a 5:1 efficiency achieved. When frequencies are freed up, governments often seek to maximum revenue through the resultant spectrum auction. This is a critical topic to address. The dominant focus especially for the LDCs should remain fixed on expanding Internet into rural areas, and not maximizing government revenue via auctioning spectrum. The degree to which the MNOs must pay extraordinarily high prices for their spectrum, is the degree to which any rural expansion will be hampered. Issues needing attention are the potential for setting aside the use of unlicensed spectrum, shared spectrum, possibly even a "use it or lose it" approach, a license provision for rural build-out requirements, etc.

- **Lowering ICT-related taxes to help expand affordability**—Most countries have moved towards privatization of their telecommunications industry, but continue to tax services and devices. Taxes should be minimized to make the services and devices more affordable. Greater affordability results in expanded adoption and with this, greater socioeconomic value results. This value will be greater than the short-term revenue gained through taxing frequencies and devices. One initiative referenced earlier that is currently underway to support this dynamic is the Alliance for Affordable Internet (A4AI).

- **New-updated and active Universal Service Fund (USF)** — A key factor limiting rural expansion of affordable broadband access is capital to build the infrastructure. Additionally, there is the need to build a rural user population that warrants the capital investment and covers the operating costs. It is not uncommon that countries set up USFs. These are typically a tax on the connectivity provider that is then used to fund build-out and subsidize lower income, lower density rural users.

While studies show many USFs are not effective, there are also successful examples and solutions, with new USFs being established even today. The key here is to have the Digitization vision, strategic and tactical plans in place to guide the targeted use of the USFs.

This needs broad public and private sector support, with USFs adopting and incorporating best practices. Another consideration is to set aside some funding from the USFs as an investment fund where a portion of the USF provides concessionary loans. These loans support build-out via low-interest loans that are repaid.

...experiences and observations from the author

Over the years there has been ongoing controversy as to the effectiveness of USFs—and for good reason. Studies by the ITU and the GSMA show there is a considerable amount of collected funds not being used (US$ 12-15M...likely more). That said, I have been engaged in several USF-related initiatives through USAID where there was tremendous success.

Vietnam—a newly formed USF was undertaken in early-mid-2000, where we provided technical assistance through the Last Mile Initiative (LMI), and moved US$ 54M out the door within the first year. The latest data shows this is now well over US$100M every year. Further, a portion of these funds, on the order of 20%, is set aside as an investment fund to support concessionary loans, not funding tenders via subsidies.

Colombia—their USF was in place and considered one of the best on our planet, though needing to be upgraded and reshaped to not only address their national backbone build-out, but to expand to include rural distribution and access.

Peru—their FITEL organization is one of the longest-standing USFs around, but Peru wanted to learn from others and make needed adjustments. It also wanted to place more focus on the non-connectivity pieces of the broader Digitization ecosystem, specifically on the demand side.

Indonesia—during 2015-2016, the Global Broadband & Innovation (GBI) program worked with the Indonesian government to: 1) build a national broadband plan, and 2) use this plan as a roadmap and architecture for reactivating a then-frozen USF fund. The plan provided the needed strategic and tactical focus for use of these funds. Their USF fund collects, and now disburses over US$180M/year.

Regional and Sub-Regional Seminars—the GBI program also partnered with Intel, where a theme was put forward that the essential requirement for a successful Universal Service Fund was a National Broadband (think Digitization) Plan developed in full participation across both public and private sectors. This theme was captured and presented jointly by USAID and Intel through a series of over a dozen regional and sub-regional seminars.

DtD Component:
Ubiquitous and Affordable Access

SDG Target 9c places a priority on expanding affordable Internet in the LDCs, though it is without specifics on "how to" achieve the target results. In many ways, this highlights the ultimate challenge for both the public and private sectors. Where earlier components predominantly give attention to the public sector's contribution to achieve the result, this component focuses more on the private sector.

> **Ubiquitous and Affordable Access:**
>
> The essential physical starting point for Digitization is the availability and affordability of Internet infrastructure and access—especially in rural areas.

This dynamic can perhaps best be described as addressing the paradox where: 1) the rural populations in the LDCs are not viewed as a viable Internet marketplace warranting private investment and associated risks; yet 2) there is an increasing global recognition that the inability to access affordable Internet is a barrier to socioeconomic growth.

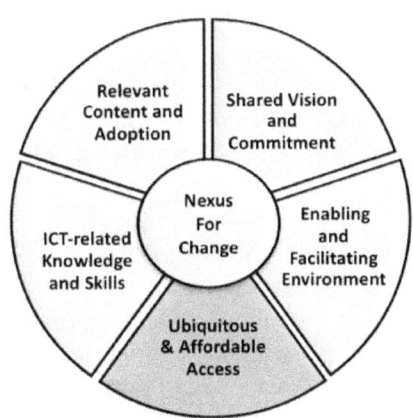

There are growing numbers of success stories that provide evidence this paradox can be, and in some locations, is being successfully addressed. The changing dynamics are from an intersection of several elements: 1) new low-cost, low-power, even off-grid technologies; 2) new successful business-financial model; 3) international private sector firms actively committed and engaged in this space; and 4) the international development community placing greater attention and resources in this arena. See a recent report funded by USAID and developed by SSG-Advisors under contract to FHI360's mSTAR Project.[55]

The priority at this stage is to engage international and local firms in ways that build off these emerging successes such that they can be massively replicated in other countries and regions.

---

[55] https://mstarproject.files.wordpress.com/2016/05/business-models-for-the-last-billion.pdf

Elaboration of the Ubiquitous and Affordable Access Key Elements:

- **Sustainable rural broadband build-out**—This element focuses on the physical infrastructure. With the Internet, this includes shared access to both domestic and international backbone. It also includes the need to expand and share backhaul capacity for the rural areas. Voice traffic can be handled via lower capacity infrastructure (i.e., satellite and microwave), where a single MNO invests in their private backhaul. However, broadband requires substantially greater capacity that necessitates a higher level of capital investment. Fiber capacity is often required. For extending broadband into rural areas of LDCs, individual MNO investment in fiber is typically not justified. Instead there is the need for a shared investment approach that provides access by MNOs, ISPs, and others.

- **Shared carrier access to international and national broadband backbone**—Having a shared high capacity backbone for low density, low-income rural areas of LDCs is the only economically viable approach. The paradox is that, on the one hand, there should not be a monopoly provider anywhere in the value chain such that it extracts monopoly pricing; yet, on the other hand, there is the need for sharing a single backhaul due to high costs associated with building the fiber backhaul. To do this requires both physical as well as regulatory accommodations. The expansion of undersea fiber in recent years is staggering, though in several countries there is but a single landing, operated by a single local operator, extracting monopoly pricing from their competitors. Having more than one cable landing can be a partial solution, as is pricing regulation where only a single landing exists.

- **Adoption of low-cost low-power connectivity solutions**—The need for electricity is one of the challenges associated with extending broadband. This is especially the situation in the distribution from the fiber backhaul into the community for reaching individuals, households, businesses, schools, etc. Fortunately, there has been a range of technology advances in recent years that significantly lowers the power requirements, while at the same time allows for local production of electricity via low cost solar solutions, wind, etc. A recent report issued by DIAL moves this topic forward. This report was prepared in February 2017 by SSG-Advisors.[56]

---

[56] https://www.usaid.gov/sites/default/files/documents/15396/Connecting_the_Next_Four_Billion-20170221_FINAL.pdf

- **Support for new and innovative business models for expanding rural broadband**—The telcom industry was initially designed for supporting voice communications where the value of the network increases as more customers come onto the network. There is a significant limitation to this model within the context of the LDCs, especially with regards to the Internet. These limitations include: 1) lower income, lower population density spread over a larger landmass, and thus lower potential revenue; and 2) higher levels of investments required.

In addition to the lower cost and higher capacity advances in recent years (e.g., 4G/LTE, Wi-Fi, TV White Space, LEOs, etc.), there are now more effective business models emerging (e.g., Mobile Virtual Network Operators (MVNOs), Mobile Virtual Network Enablers (MVNEs), Wireless Internet Service Providers (WISPs), etc.).

...experiences and observations from the author

Drilling down a little further on our engagement in rural **Mongolia** in the mid-2000s, we ran across a couple situations that were totally unanticipated. Yet they turned out to be fundamental to the success of providing sustainable rural telecom. This was in the rural voice arena, not with the Internet as the primary driver.

Who Would They Call? —during the initial trials of the low-cost Voice-over-IP rural networks (using open source software), the owner of Incomnet wanted to experiment by allowing free calls for the first couple days. The idea was to observe whom the people living in these rural soums would call. So, there was no cost for any call for two days. The findings: The people called all over our planet, including Singapore, South Korea, the U.S., Italy, France, Japan, among other places. We had anticipated local calls within the soum and within Mongolia. Interestingly, even after charges were put into place, the global calls continued.

Financial Support from Across the Globe—once friends and relatives knew the locals had a phone in their ger/yurt, the calls started coming into the soum. And with these calls came call termination fees that in part, helped pay for the local rural network. It must be assumed that those living outside the soum had higher income levels to where this wasn't an issue. By all accounts, this has happened in other rural networks as well, and it is clearly something to factor into the financial modeling for rural deployments.

- **Open and competitive telecommunications market**—The history of telecommunications repeatedly proves that in country after country competition is effective in expanding access and making it more affordable.

- **Local IXPs to reduce international transit for local use of local content**—with the Internet, another component growing in importance is local content. There is also the need to keep local traffic local, where a central Internet eXchange Point (IXP) routes traffic locally, and in doing so, reduces costs associated with international transit.

- **Rural community access centers**—While there is a somewhat failed history of telecenters dating back 15-20 years, there are also successful examples. These had a collective approach for a cost-effective way to expand low-cost local access. A rich resource for this space is Telecenter.org[57]. Another recent resource is from Francisco Proenza. In 2015, he completed an extensive study and update on public access, capturing his research and findings in a report titled, "Public Access ICT Across Cultures: Diversifying Participation in the Network Society.[58]

- **PC-Smart Phone purchasing programs**—Over the years there has been a range of successful programs to support individual purchase of PCs through MNOs and other organizations. More recently, it is smart phones. Typically, the MNO's provide incentives and options for purchase, though this remains a challenge for those living in the rural areas of LDCs.

---

[57] http://www.telecentre.org

[58] https://idl-bnc-idrc.dspacedirect.org/bitstream/handle/10625/54174/IDL-54174.pdf?sequence=1&isAllowed=y

DtD Component:
ICT-Related Knowledge and Skills

Many countries with the greatest need for leveraging ICTs for socioeconomic growth often lack a strong local ICT sector. International attention to this issue is starting to take place.

> **ICT-Related Knowledge and Skills:**
>
> The future of Digitization will be in the hands of the country's youth. This requires creating opportunities and making investments for expanding the local ICT-related skill base.

There are growing numbers of successes with models and lessons learned to replicate and scale by lesser-developed countries. ICT-related knowledge and skills leads to higher paying jobs in the local economy that most often attracts youth.

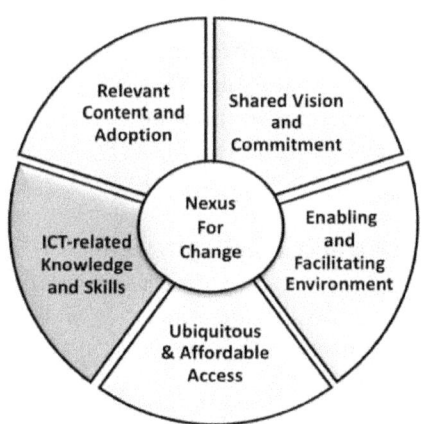

High-tech firms such as Intel, Cisco, Microsoft and others have developed rich skill-building programs with curriculums to support the local adoption of their products and services. This includes online and local training of specific skills (i.e. Cisco Academies), as well as access to in-country and regional certification testing facilities.

Elaboration of the ICT-related Knowledge and Skills Key Elements:

- **ICT job skills development**—A fundamental cornerstone of digitization is having a local indigenous ICT skill base. This is frequently non-existent for the LDCs. Too often, both past and present, the international development agencies focus only on the adoption, use, and the technical support for their specific project. Additionally, the private sector views the opportunity as an export of products and services. Neither provides the needed local dynamic. ICT skill-building must be broader and deeper, and be included in the National Vision and Planning process.

- **Directly engage international firms such as Cisco, Microsoft, Intel, IBM, HP, Google, Facebook, etc.**—Virtually all international high-tech firms build and continue to refine ICT skills development materials, classes, and on-line courses. These are aimed at skills needed to support their respective products and services, but they also include skills and knowledge that are transferable.

- **Technology training and certification programs**—Most high-tech companies provide formal testing and certification for professional-level skills. This ensures a complete solution-set by the vendors and in doing so, extends technical skill sets into the local populations.

- **ICT-related user skills training**—In addition to programs for developing professional skills, several technology firms have invested in user-focused training that targets the adoption and use of their products and services. This too is a critically needed component of Digitization.

- **Innovation and entrepreneurial programs and laboratories**—In recent years there has been international support for establishing local in-country ICT-oriented programs. Often the programs are connected with local universities, but some are stand-alone and support a broader base of participants.

- **Local university ICT-related certification and degree programs**—ICT degreed programs is another essential component for building a strong local ICT industry. These can be undergraduate and graduate level, as well as certification programs that support university students majoring in other areas such as business, management, economics, etc.

- **International university linkages programs**—One approach to advancing ICT-related training is through collaborative efforts between universities, whereby students can temporarily relocate to an overseas university for educational opportunities not available locally. Programs may also offer teaching opportunities for university professors to travel internationally and teach at a local university.

- **Building-Enhancing the local ICT Industry**—Long-term sustainable success needs to reach beyond the individual level of building technical skills, to include establishing a facilitating business environment for building local high-tech businesses. This industry can provide the essential support for expanding both the public and private sector's adoption of ICTs, as well as contributing to the local economy.

...experiences and observations from the author

USAID, through a range of its ICT-related initiatives, has placed high priority on capacity development. The following are a few examples where I was engaged.

Armenia—USAID brought the Cisco Academy into three State Universities, leveraged their high-tech diaspora in the U.S. to conduct in-country workshops, and later launched MSIS degrees in three State Universities.

Eritrea—USAID brought two Cisco Academies into the country, one on-campus at the University of Asmara and one-off campus. We also set up two computer labs.

Myanmar—this is another country where USAID supported the placement of four Cisco Academies within Universities—two in Yangon and two in Mandalay.

USTTI—the United States Telecommunications Training Institute is another rich source for a broad range of technology-related awareness and skill building. Through the USTTI foreign individuals are brought to the U.S. for a wide range of training opportunities—many hosted by U.S. based high-tech firms. Through the LMI program, USAID partnered with Intel and USTTI in designing a two-week seminar hosted by USAID and Intel for advancing many country engagements in ICTs. Seminars were conducted twice a year over a 3-year timeframe.

Cisco Networking Academy Source: https://www.netacad.com
USTTI Source: http://ustti.org

...experiences and observations from the author

In the country of **Armenia**, our engagements started in 2000, where USAID supported a range of ICT-related projects for well over 5 years.

These projects cut across several key Digitization components as put forth in this book. The ecosystem of priorities was shaped though the development of a National ICT Plan with support and joint engagement by USAID, the World Bank, and the Government of Armenia.

Technical skill building was critical as was the establishment of MSIS degree programs in three state universities.

This was complemented on the demand side with building the local high-tech industry. The foundation for this focus was the recognition that Armenia was once a center of IT innovation during the former USSR period. To rebuild the needed local high-tech industry, our project leveraged the Armenian diaspora in the U.S. through an agreement with ArmenTech. Under this arrangement we funded travel-related costs for executives and technical specialists to come and host a range of workshops and training sessions in Armenia. The arrangement also allowed for some Armenians to travel to the U.S. and do internships in various companies. Other USAID-funded programs supported the strengthening of the local ICT sector.

A recent report by TNW indicates that the Armenia tech industry is now growing at a rate of 20%—exceeding the country's 2% economic growth. There are an estimated 400 IT companies now in Armenia, forecasted to total US$ 475M by 2018. U.S.-based firms such as Intel, Microsoft, Google and Oracle are all present in Armenia.

Sources:

https://thenextweb.com/asia/2017/03/17/armenias-rising-tech-scene-new-silicon-valley-former-soviet-union/#.tnw_qsN3uz57

https://www.evnreport.com/economy/armenia-s-tech-sector

DtD Component:
Relevant Content and Adoption

While the Internet access gap draws the most attention (as reflected in the earlier ITU graphics in Chapters 2 and 4), the Internet statistics only serve as a proxy indicator. Ultimately, it is the application and adoption of technologies that delivers the sought-after socioeconomic results.

This is critically important within the LDC context, as this is where the demand is most often at a very low level. It is also where there is the greatest need for a focused, aggressive, integrated, and sustained series of initiatives that produce impact.

> **Relevant Content and Adoption:**
>
> Ultimately Digitization delivers its socioeconomic value via broad based adoption of ICTs. In the LDCs, the government plays an essential kick-starter role via providing a range of online social and economic value.

Here the role of the national government is essential. And this role is not simply to provide the enabling and facilitating environment for infrastructure. Rather, it must include building content and services where the government is actively engaged in delivering online socioeconomic-related content and services. This includes not only providing citizen and business services as part of a broad e-Government initiative, but also providing educational services, agricultural services, health services, written language skills, etc.

Another national government priority includes working with the private sector and local governments to provide local digital literacy training, community-level access centers, rich content, services, and applications that meet local needs.

Elaboration of the Relevant Content and Adoption Key Elements:

- **Adoption of ICT by local governments and businesses**—There is an interesting dichotomy between the developed and the least developed countries. In the developed countries, the private sector was the first demand-driver for ICT adoption, and the public sector was relatively slow in the adoption of ICTs. In the LDCs, the model must be reversed. This is because the private sector is most often not positioned to take the lead, and the government is—should the government adopt a vision and plans to do so. The government also has the advantage of adding often-needed government transparency for its citizens.

...experiences and observations from the author

To a significant degree, the economic challenge of telecom is aggregating demand sufficient to capture revenue that will cover the return on the initial investment and operating costs. A couple initiatives, one out of USAID's LMI and one out of the GBI, addressed this topic where Education was the dominant driver.

Macedonia LMI—this may well have been USAID's largest success where demand aggregation was the essential driver for connecting over 450 schools to the Internet. The project focused on establishing broadband, placing PCs in schools, creating student curriculum, and building teacher capacity. USAID/Macedonia issued a Request for Proposal (RFP) for provisioning the schools with broadband, with the understanding the USAID Mission would pay for this capacity for two years at an established rate. The ISP winning the contract leveraged the committed income stream to finance build-out of the national network.

Jamaica GBI—this initiative was much smaller than the Macedonia LMI, but again focused on connecting schools with broadband. This too was in support of a USAID educational initiative in partnership with Microsoft and a local carrier, FLOW, and in support of Jamaica's Vision 2030. While schools provided the rural community point-of-focus, as the networks were built, others sought to come online—adding to the overall rural community demand. The funding was via the partners as well as from Jamaica's USF. The technologies were TV White Space and Wi-Fi.

- **User-Focused Digital Literacies**—In addition to developing and expanding local technical skills, it is imperative to build and enhance user-oriented skills. Ultimately the local school system can deliver this training, but initially it may need support through a separate initiative.

> *...experiences and observations from the author*
>
> In **Vietnam**, parallel to USAID's engagement in operationalizing their USF and deploying two rural WiMAX networks, we partnered with Microsoft, Qualcomm, Hewlett-Packard, EVN Telecom and the Ministry of Education and Training (MoET), to launch "Training Online Programs and Incubation for Communities (TOPIC64)". Pham Minh Tuan at the Hanoi University of Technology's Center locally managed the project for the Research and Consulting Center (CRC).
>
> The curriculum was built around Microsoft's Unlimited Potential Program. During 2006-2007, the initiative established 64 centers, one in the capital of each Province. 10,466 students and 474 teachers were trained in digital literacy through these Centers. An additional 70,692 students were reached through 426 affiliate Centers.
>
> This effort was based on a sustainable business model that continues to this day — 10 years later. In addition, Tuan at the CRC expanded upon the on-line education model and launched TOPICA, which provides online English language and college degree programs. TOPICA has since expanded to other Asian counties.
>
> Sources:
>
> https://evolllution.com/author/tuan-minh-pham/
>
> https://www.dealstreetasia.com/stories/vietnam-based-edtech-platform-topica-takes-over-incubator-hub-it-26494/

- **Support for local ICT sector and ICT-enhanced services imbedded into local sectors**—The Digitization model includes a focus on the adoption of ICTs by various sectors, with the recognition that the value derived from this adoption varies considerably sector-to-sector. To maximize the potential gain of Digitization, the focus must not only include promoting the adoption of technologies, but also placing a national priority on developing industries that can benefit the most from this investment.

- **ICT value-add focused on supporting and achieving Sustainable Development Goals (SDGs)**—Another strategic opportunity for the adoption of Digitization is placing a priority on those SDGs pursued by the country itself, most often with support from the international development community. One of the valuable resources developed in 2015 was the "SDG ICT Playbook: From Innovation to Impact." This is an excellent resource and is available at NetHope's Solution Center.[59]

- **Expanding use of applications and content for social and economic value**—An essential component of leveraging the Internet is gaining access to locally relevant information and services. This contributes to advancing the socioeconomic positioning of the country, in both the public and private sectors. The vision, strategic and tactical plans should incorporate components that move this agenda forward.

- **Develop targeted applications and content for advancing public-sector services such as government, education, health, and agriculture, as well as private sector services such as agriculture, financial services, and trade**—In the LDCs, the government is in a unique position to take the lead and provide the kickstart for Digitization. This is due in part to the nature of the services it provides to its citizens, as reflected in the title of this element. Digitization can be highly leveraged to add valuable services to a larger population. This is especially true when most of the citizens live in rural areas and are otherwise not reached. This is also where the international development community via International Non-Government Organizations (INGOs) provide specific value-added services. A key emphasis is to ensure digital services are built once, placed on the cloud, and accessible to all.

---

[59] http://solutionscenter.nethope.org/toolkit/view/sdg-ict-playbook-from-innovation-to-impact

...experiences and observations from the author

One of the principal areas of our work in **Armenia** was the financial sector. This work took place in the early-mid 2000s before mobile money gained attention. Our efforts focused on the fundamentals of connecting and integrating Armenia to the rest of the world's economy via financial networks, as well as enhancing local financial services. This included the following:

SWIFT—our support for Armenia's financial sector started with bringing international funds transfers into Armenia through the Central Bank of Armenia (CBA).

FinA—this project included the development of a financial auditing package for the CBA, where they could have real-time oversight and auditing of the banking industry. This was Open Source and has since been put into production by several other countries as well. See: https://sourceforge.net/projects/fina/

CBA Net—with CBA's linkage with SWIFT, the CBA Net connected the CBA with all the then seven (7) commercial banks operating in Armenia. This allowed real time-sharing of banking data, daily clearing of financial transactions, etc.

ArCa—the USAID project introduced Armenian Card (ArCa) branded debit cards though all the commercial banks. This included establishing a single clearing house for transactions, working with employers to make labor payments into employee ArCa accounts, as well as working with local merchants to accept payments via the ArCa debit card.

ATMs—at the time cash was still king and ATMs were non-existent. To allow ArCa customers to get cash out of their debit accounts, the project put approximately a dozen ATMs across Yerevan. Initially this was a high demand that soon lowered as the acceptance of the ArCa debit card by local merchants caught on.

Online Payments—the project allowed for making utility payments via the ATMs, with the design for future payments to be made via mobile phones.

Credit Cards—after the ArCa debit cards were in place for two years, the project brought in both VISA and MasterCard to work with the local banks to introduce locally issued credit cards. With ArCa cardholders now having a financial transaction history, credit limits could be established with minimal risk to the banks.

Foundation for Mobile Money—USAID's support for the core backbone services across the financial sector in Armenia formed the foundation upon which to introduce mobile money. This was a logical and simple extension of the ArCa debit card. This was a unique experience supported by the local banking industry rather than a MNO implementation.

- **Promote mobile applications and local mobile application marketplace—** We are well past the era where the personal computer is the dominant user device. In sheer numbers, the smart phone has surpassed the personal computers and laptops. A focus on cloud-based mobile access maximizes individual adoption as the networks extend into the rural communities.

> *...experiences and observations from the author*
>
> One of the areas of periodic controversy is referred to as zero-rating. Most often this term relates to Facebook's Free Basics approach of working with carriers to provide no cost access to selected sets of information. Somehow this gets interpreted as unfair competition, with several countries prohibiting the practice. In my opinion, this is simply an extremely limited view. Take out Facebook for a second.
>
> The all-too-often situation, especially in lower income countries, is relatively straightforward as to advancing Internet use: 1) there is a cost issue. Internet access is often unaffordable; 2) there is the issue of needing to stimulate written language and digital literacies; and 3) there is the need to fuel and kickstart use through expanding demand by placing information and value-added services online and available as an essential element.
>
> The actions by Facebook are not an end in itself, but rather a catalyst—a catalyst through which all will ultimately benefit by getting people online. Perhaps there are better approaches, and if so, others should step up and implement alternative approaches, rather that attempting to stop at least a partial, interim solution that moves things in the right direction.
>
> In my view, it is simple: stopping Free Basics should only be done under one situation—where those stopping its introduction offer a better solution to where more people are brought online sooner and at a lower cost, and with access to more information. It should not slow things down, but rather, speed things up. Taking a "no" position without launching a better solution-set simply works against advancing the very core elements that are most needed.

DtD: Summary-Conclusion

The DtD model places a high priority on two critical components that too often do not get sufficient attention.

- The absolute need for putting into place an active government-led "Nexus for Change" with support from the highest levels in the government, the private sector, and the international development community. This is simply not an option if the issues are to be successfully addressed in a timely manner.

- The LDC governments must play a critical role in delivering a range of value-added government social and economic-related services as the cornerstone for addressing the demand side of the equation. This serves as an essential kickstart that makes up for the missing private sector demand-related content and services in these economies.

Together, these two components serve as the foundation for advancing the socioeconomic gains much needed by the LDCs and afforded through Digitization.

While SDG Target 9c emphasizes expanding affordable Internet access in the Least Developed Countries by 2020, connectivity is but one component of the broader scope of Digitization. To achieve and even accelerate success in the LDCs, it is essential that we move well beyond the technologies and focus on their applications—where the value-add is achieved by delivering measurable social and economic gain.

If this Digitization approach is not aggressively pursued, the current gap between the Developed and Developing countries and the Least Developed Countries (LDCs) will continue to expand throughout the 2015-2030 timeframe of the SDGs.

CHAPTER 7

Sharpening the Focus for the Next 15 Years

During the MDG timeframe there were key dynamics taking place in the international development arena to expand affordable connectivity and leverage its socioeconomic value-add. As this book is being written, the SDGs are now in place charting the direction for the international development community through 2030.

This final chapter concludes by looking forward to the next 15 years of the SDGs and on the possible dynamics associated with Target 9c:

> *"Significantly increase access to information and communication technology, and strive to provide universal and affordable access to the Internet in the least developed countries by 2020."*

Without discounting the importance of this single statement, it is instructive to note that the operative word relative to access in the LDCs is rather weak—"strive." And while the target contains a date of, "by 2020," even as early as 2016, the broadly accepted position by the international community was that this target will not even be met by 2030—10 years later and the final year of the SDGs. The target also focuses only on providing "increasing access" and "affordable access" to the Internet, whereas the challenge reaches well beyond access. If the whole of the Digitization ecosystem isn't addressed in an integrated manner, the sought-after socioeconomic benefits derived through the focus of this SDG's Target will simply not be realized.

The Digitization model put forth in Chapter 5 was constructed and refined to directly deal with the essential requirements to address the broader ecosystem.

This chapter puts forth several topics that must receive attention and prioritization in order to move from the model to achieving on-the-ground success.

1. Establish a Global-Wide Strategic Focus on LDCs

The international development community, along with the public and private sectors, have reached several conclusions:

a. An increase in Digitization results in increased socioeconomic development;

b. LDCs are at the lowest level;

c. The LDCs continue to fall further behind relative to other developing countries;

d. At this stage there is collective awareness, knowledge, and technical solutions sufficient to address this situation; and

e. A limited number of businesses are engaged in this area to where successful business models are beginning to surface showing the associated issues.

The current challenge comes down to implementing a joint agenda between the international and national levels of government and the private high-tech sector. It requires a collective, focused and resourced agenda to address the challenge. The SDG Target is but a first step, and far from sufficient. Turning the issue around a bit, it can perhaps be stated better as a question, "When it comes to Digitization, and knowing what we know, why aren't LDCs a high priority across the international development community?"

Needed Refocus: We must move from seminars, reports, and discussions, to massive wide-scale action across our planet. As Nike stated some decades back, the needed reform is "Just Do It!" Yes, details must be worked out in the design and execution. And, yes, there are associated issues to address. But this is a doable LDC-related challenge that we can address. It is similar to what was undertaken in the rural U.S. a hundred years ago to extend rural telephony, and there are lessons to be borrowed. This includes governments not being an obstacle but ideally, contributing value-add in the mix; the availability of off-the-shelf technology that can be deployed in a parallel fashion across thousands of rural locations (most logically WISPs but also 4G/LTE and soon 5G small cells); and the expansion of locally-relevant government and private sector value-added content and services (with government-provided services such as education, health, etc., being a critical starting point).

Actionable: Establish a collective prioritized international agenda on Digitization for the LDCs—such that it is a strategic focus of both the development community and the countries themselves. International donors should establish new and separate Digitization programs that sit alongside exiting programs such as Education, Health, Agriculture, Environment, etc.

2. Shift the International Focus:

During the 2000-2015 MDG timeframe, the international development community focused on three dominant themes:

a. awareness building and advocacy for ICTs;

b. developing and promoting best practices; and

c. expanding affordable access to the Internet.

Tremendous progress occurred in expanding the Internet; however, there remains the yet-to-be-realized challenge of the LDCs.

Needed Refocus: Looking forward to the next 15 years guided by the SDGs, we are well past the awareness and advocacy phase. The essential focus as we look towards 2030, in a word is, "Doing." The awareness is near universal, the technologies are available and increasingly affordable, and expanding each day. And there is a growing inventory of workable business models capable of achieving sustainable success in the LDCs. The essential missing keys are: 1) a refocused international agenda that supports action; and 2) the need for the countries themselves to step into a position of ownership and in-country leadership. Without a substantially more aggressive approach, the default path is that the LDCs will continue to fall even further behind.

Actionable: Pursue a more focused global "doing" agenda. The multilateral organizations, along with bilateral donors should more actively support on-the-ground programs.

3. Establishing a Solid National Foundation

It is not enough to advocate for and to begin working in the LDCs. It is imperative to specifically focus beyond larger urban centers to reach the rural population. Digitization takes investment and investors seek a return on their investments. Addressing the risks associated with the higher-per-user cost factor and an initial lower adoption level presents a constraint. The default is that private sector investments will not be made—the current reality. This constraint must be addressed.

Needed Refocus: For all the promise of Digitization, it must be built on a solid and stable national platform—a platform with a high national priority. Socioeconomic development is the sought-after result of Digitization. Those living in rural communities are often forgotten, yet they are the ones who stand to benefit the most. Therefore, there is simply no substitute for making this a national priority.

Actionable: LDC country governments must elevate Digitization to the level of a national strategic focus and support this with plans and action.

4. Place a High Priority on Digitization

Internet growth over the timeframe of the MDGs (2000-2015) reflects that whatever approach powered the rapid growth of the Internet in the Developing and Developed countries, did not take place in most LDCs. Even with the SDG Target 9c in place since 2015, it is still not moving towards the desired target impact. Nor is it forecast to achieve the target to any significant degree for the next decade plus.

Needed Refocus: The international development community, along with the governments of the LDCs must rethink, partner, and act on several key elements. They need to mainstream digitization into their international development agendas. This would include providing support for:

a. Establishing a facilitating environment;

b. Lowering investment risks of corruption and instability;

c. Removing the domination of monopoly providers that exist along the Internet value-chain that extract monopoly pricing;

d. Lowering the costs of delivery by sharing passive and active infrastructure, adopting newer-lower cost technologies and adopting emerging new business models;

e. Building new local hosting and locally-relevant data sources;

f. Focusing on expanding written language, reading, and computer skills;

g. Putting into place adequate skilled technical resources; and

h. Promoting and engaging governments and private sector businesses to take advantage of leveraging ICTs/Digitization within their respective areas of responsibility; and, providing citizens support by improving business operations.

Actionable: The international community needs to consider developing a prioritized engagement agenda for addressing the unique situations that are present within the community of those countries classified as LDCs. A potential model may be taking shape by the European Commission, where in May 2017, they presented a strategy to mainstream digitalization into EU development policy.[60]

5. Make Digitization a National Agenda

The LDCs are disadvantaged on many fronts, as there are simply too many high priorities they face that need attention. The result is that Digitization and its various components are generally shoved into the background and viewed as something to address later. To move forward on Digitization requires national treasure. And until the national public and private sector leaders view Digitization as an essential high priority component for achieving socioeconomic growth, the essential investments will not be made.

Needed Refocus: Digitization must be viewed as a critical agent of change that contributes across the country's entire socioeconomic spectrum. The proposed "Nexus for Change" component must be recognized and pursued as an essential core. The next two priorities are also critical—establishing a "Shared Vision and Commitment" and implementing an "Enabling and Facilitating Environment." These are fundamental for success. In discussions I've had with several from the multilateral development banks, they have affirmed that the level of loan requests they receive from countries reflects that digitization is most often a very low national priority. This current stagnating position needs to change.

Actionable: Establish Digitization as a new country priority to support the development and refinement of a national agenda. The multilateral and bilateral development organizations, along with the local government and private sector, provide support. There must be specific initiatives to make the plans a reality.

6. Emerging Game-Changing Technologies

There has been a staggering level of technology advancements over the period of the MDGs. These private sector-led advances were market driven, and predominantly driven by the higher density and higher income markets. That said, in recent years newer solutions have emerged that hold promise for providing cost effective solutions to reach rural populations more typical of LDCs. Low-cost Wi-Fi, Wi-Fi Mesh, TV White Space, 3G-4G/LTE small cells, along with

---

[60] https://ec.europa.eu/europeaid/news-and-events/european-commission-presents-strategy-mainstream-digitalisation-eu-development_en

lower-cost higher-capacity long distance point-to-point microwave solutions, are part of this mix. And there is promise of 5G small cell solutions on the horizon that will be capable of providing fixed and mobile support.

Another technology dynamic currently unfolding and potentially reaching the market in the next 4-6 years is constellations of low-altitude micro-satellites. These micro-satellites hold promise of providing worldwide ubiquitous voice and broadband access to every square inch of our planet. Several initiatives are currently funded and making great strides. The combination of 5G (small cells) with these swarms of high-capacity low-altitude micro-satellites holds the potential for making affordable Internet available globally. Ubiquitous worldwide broadband will be here during the 2025-2030 timeframe. And with this, the connecting technology will become a non-constraining issue towards advancing broad scale Digitization.

Needed Refocus: By default, emerging technologies will follow the patterns of the past—they will first be deployed into the higher density, higher income environments. Pursuing the normal serial deployments driven by revenue stream must be replaced with a different model. The international development community, in partnership with high-tech firms, needs to pursue an aggressive parallel approach to target and accelerate the rollout of these technologies to rural populations in LDCs. The revenue will not be as great, but proportionately the socioeconomic impact could be significantly greater.

Actionable: Initiate a targeted agenda to take early advantage of emerging technologies—where lower cost broadband becomes available everywhere—in all countries, including the rural areas of LDCs.

...experiences and observations from the author

What is now referred to as "Small Cells" got its start in 2007 with the formation of the FemtoCell Forum. The Forum focused on filling gaps in cell coverage inside office buildings, urban centers, and cellular shadow areas in suburbs. Professor Simon Saunders initially chaired this Forum.

During this timeframe I was involved in kickstarting the implementation of small rural networks in Vietnam and Mongolia. These networks relied on VoIP servers and Wi-Fi/VoIP phones and VoIP phones. This gave us the needed low-cost switching and interconnection to the PSTN and mobile networks. It worked—sort of.

One day, Dave Lyman, a colleague, contacted me and sent me a FemtoCell to try out in my residence. I live in rural Oregon, where coverage was not great. I plugged it into an Internet cable modem and it made a huge difference! We continued to experiment, including placing a small FemtoCell off the back of a satellite. It too worked very well. A <$200 Samsung FemtoCell, on the back of a VSAT gave us a 1 mile plus radius of coverage and provided 3G voice and data services.

I reached out to Professor Saunders and described our experimentation. He invited me to speak at their annual FemtoCell Forum Summit in San Francisco in November 2010. My message to their members was pretty simple: "You don't appreciate the value of what you've developed with these small cells. You have invented a solution-set targeting higher-income, higher density urban and suburban populations in developed countries. And what you have also invented is a solution-set with the potential for connecting the most impoverished rural populations around the globe."

Ratchet forward 10 years. FemtoCell Forum has since been renamed the Small Cell Forum. And smaller Metro Cells have become standard in high density-high demand urban settings to support frequency reuse. Several firms have entered the rural marketplace deploying small cells. Further, 2.5G and 3G technologies have given way to 4G/LTE and soon 5G, where the small cell architecture will over time become dominant.

Example: in 2010 we launched an initiative in the Democratic Republic of Congo (DRC) to reach rural communities. USAID supported a demonstration of four solar-powered, satellite-based small cells. Vodacom, the local partner now has on the order of 800 such installations across rural DRC.

Over the next 5-10 years, add the next generation of constellations of satellites consisting of hundreds of micro-satellites capable of delivering broadband capacity to virtually every inch of our planet's surface. And with these dynamics coming into play, it is even more essential that our focus shift towards strengthening international and national leadership and addressing the other Digitization components to deliver the sought after socioeconomic impact to the most impoverished.

7. Promote and Adopt New Innovative Business Models

The default business model has been that large telecommunication providers, mobile network operators (MNOs), provide telecommunications services. As competition commenced and the marketplace opened to support multiple MNOs, the coverage area expanded and pricing was reduced. This was augmented by the ISPs and WISPs providing Internet.

The first bookend, the "Wireless Village: First Mile First", reflected on the telephony history in the U.S., where the government was not the monopoly provider, but rather Bell was the monopoly for 18 years, the life of the phone patent. As the patent came to an end, literally thousands of small independent rural community telcos were established to meet local community voice services. To a degree, today there is the near equivalent, where in addition to the MNO's, independent ISPs/WISPs are providing Internet.

Needed Refocus: There are alternate business models emerging that seek a shared-asset approach for both mobile and the Internet. Sharing passive infrastructure has been around for some time. This includes competitors relying on shared civil works infrastructure of conduit, towers, etc. More recently MNO's rely on independent tower companies that build the towers and lease access to the tower to multiple MNOs. However, to service small rural communities, especially in LDCs, sharing active infrastructure is needed. This is currently being done through mobile virtual network operators (MVNOs), and mobile virtual network enablers (MVNEs).

Actionable: Support broader adoption of these models, where a single network operator provides both voice and Internet though a single network in support of multiple operators. This will extend the reach of voice and Internet into rural areas of LDCs faster and in a much more cost-effective manner. In locations where local WISPs distribute Internet, close collaboration and reliance of shared backhaul infrastructure is also needed.

8. Shift from Donor-Led to Private-Sector Led Partnerships

For the international development-donor community, the nature of providing support in the space of connectivity is unique from virtually any other sector program. Here there is the opportunity to form rich donor-public-private partnerships, where each contributes to a broader solution from their respective strengths. These joint interventions can be relatively short, with the engagement putting into motion a trajectory where the private sector continues expansion beyond the life of the partnership. This donor-private partnership model started to take shape in the mid-late MDG timeframe.

To a degree, the emerging business model for many high-tech firms is based on expanding client-service-based revenue stream. This is dependent on growing the number of people accessing their services and increasing their intensity of use. This business model encourages firms to make infrastructure investment, be it expanding broadband access via a range of wireless technologies (including Wi-Fi, new 5G small cells), or space-based broadband solutions via low altitude balloons, drones, or swarms of microsatellites.

Needed Refocus: The default pattern has been international donors taking the lead contributing legal and regulatory support, with the high-tech private sector firms providing technology solutions. Today, the high-tech private sector is positioned to change the nature of public-private partnerships and take the lead.

Actionable: Under the new scenario, the private sector takes the lead, and the public-donor community provides targeted value-add in areas where they can provide strength in policy-regulatory changes and increased local human capacity. This new model holds untapped potential, but it will require more concentration and refinement as we move forward.

9. Building Local Capacity and Skills

As reflected in the Digitization model, strong local capacity and skills are needed to support the commitment to Digitization. Building these skills can be achieved through a wide spectrum of sources, including equipment-software providers, certification programs, entrepreneur initiatives, online access, as part of establishing a technology center or hub, or from universities, and trade schools. Often support from the international donor community is limited to the donors' development initiatives. Local staffs are trained as part of a donor project requiring ICT-related skills and support. Under this scenario it is not uncommon that when the project is completed, and the donor support comes to an end, these skilled workers move on to higher paying jobs in the private sector—with the donor-funded project at times placed in jeopardy.

Needed Refocus: Building local ICT-related capacity and skills should be addressed as a separate initiative, and not just imbedded in a specific ICT-related donor project. Advancing local skills should be viewed as building a critical national resource pool upon which the country-at-large will benefit. It is also a value-add in that this initiative will most often train and support the youth, and can be used for advancing gender equality.

Actionable: There must be a focus on developing technical, analytical, and soft-skills that are essential for reengineering processes and procedures to extract efficiencies from applying technology—essentially, building or strengthening a local tech industry.

10. Digitization of Government Services

It is imperative that the national governments of LDCs play a leading role in the demand component of the Digitization ecosystem. The LDCs typically don't have strong private-led commercial markets, nor are there large firms and/or industries to drive any significant adoption. Therefore, the governments must play this critical role. This can be achieved by adding and extending a wide range of Internet-based citizen-business services.

The government is in a unique position to provide value-added services to the entire population. It can effectively aggregate demand across society for services like education, health, and agriculture, as well as a wide-range of government services such as land records, birth records, and other government services. Every citizen is a potential "customer." This kickstart role holds a unique potential in setting the stage for added uptake by individuals, the government, private sector, and entrepreneurs.

Needed Refocus: In the more developed countries, the private sector has led the way to adoption of ICTs, whereas the government has lagged. This was especially the case in the migration to leverage the Internet. For the LDCs, there is both an opportunity to improve access to basic services to those in the more rural communities, and to establish a national consistency for delivering services.

Actionable: National governments should focus on the development, expansion, and adoption of the Internet. Governments can kickstart the provision of key services, and in doing so, support an aggregated demand that serves as a catalyst for building out the national network. With this put into place, the private sector will follow, adding further content and services—all of which contribute to the socioeconomic value obtained through Digitization.

11. Targeting Private Sector Value-Added Priorities

The LDCs are at the lower end of countries benefiting from Digitization. Raul Katz's writings reflect two key themes that relate directly to LDCs. Those themes are the adoption issue where less people have access, and secondly, the nature of industries located in the LDCs. For example, Katz's data reveals there is a greater impact realized from Digitization of financial services than from agriculture. Yet in the LDCs, it is agriculture that is generally the dominant sector.

Needed Reform: As countries pursue a Digitization agenda, it is imperative that it is driven by a national agenda and priorities, and not by the donor community. This creates a rich opportunity within LDCs for donors to provide experience, expertise, and funding support, and where the local governments can step up to ensure that the multilateral and bilateral support is well coordinated and managed.

Actionable: The LDCs must actively develop and manage a National Digitization Plan to guide and track their implementation priorities. The donor community will support many of the projects, along with support from the private sector. The Plan helps to ensure an integrated whole. This Plan should prioritize initiatives that provide the highest return on the Digitization investment possible within the LDC setting. This needed refocus reflects the value of Development as the primary focus and Digitization the secondary focus: "Development through Digitization".

Concluding Thoughts

My final thoughts pull together key themes needing prioritized attention by the international community, where both the public and private sectors seek to leverage digitization for socioeconomic development in the Least Developed Countries (LDCs).

The ITU's Internet user data for the LDCs during the MDG/WSIS timeframe (2000-2015) reflects that development and marketplace dynamics are simply not working for these countries. And while Target 9c of the SDGs mentions Internet access for LDCs, with a target date of 2020, the ITU's current forecast indicates the lack of progress will continue.

Change must happen. Substantive change. Out-of-the-box, order-of-magnitude change. The business model of "build it (telecommunications) and they will come," typically does work in more advanced economies, even in some of the mid-upper tier developing countries. This is possible because their economies are sufficiently large and developed to the point that there are latent demands warranting the build-out investment. For mobile telephony, this has even worked in LDCs except for reaching those living on the rural edge. But with few exceptions, it is not working for expanding affordable Internet in the LDCs rural areas.

An essential outtake of exploring the expansion of affordable Internet in the LDCs is that this issue cannot be addressed only from the supply side. Rather, it must be addressed through a comprehensive approach that also addresses the demand side of the equation. And the LDC governments have a lead role to play on both the supply and demand.

Paradoxically, the countries least well-positioned to pursue integration of the components put forward in the DtD model, are also those countries most in need of socioeconomic development—where digitization holds promise for making a significant socioeconomic contribution.

Another unfortunate paradox is that the international development agencies from the most technologically advanced countries—those countries that have benefitted the most from digitization, as a rule, have only marginally focused on this topic during the period of the MDGs. Nor are they currently aggressively pursuing this under the newer SDGs.

Speaking from a position of passion, as well as from a position where I have had the opportunity to participate first-hand in this space, the global commitment relative to the SDG 2015-2030 timeframe has not yet adequately materialized to address this overwhelming need. Ideally, the development community should adopt Digitization as a mainstream program to where it exists alongside existing programs for health, education, agriculture, etc. These new Digitization programs must include a concentrated resource commitment for expanding and leveraging the Internet in the LDCs. It is only then that these countries will reap the socioeconomic impacts afforded through Digitization. This is currently an unmet challenge.

As we look towards 2030, we must collectively focus on building the national level government priority in LDCs. This should include charting their direction, expanding value-added government-related services, and moving as quickly as possible to develop a range of globally-adaptable technology and business models that provide rural Internet deployments.

Today we are positioned to:

1. Capture the dynamics, observations, and lessons learned from the prior 15 years of the MDGs where ICTs, specifically the Internet, has played an increasingly important role in global socioeconomic development; and

2. Explore moving forward into the next 15 years of the SDGs, where we can leverage lessons learned and successes with a focus on adaptation to the LDC settings. We must sharpen our focus on leveraging the new emerging technologies and innovative approaches that will have greater socioeconomic impact in a shorter timeframe. This needs to be pursued in a massively parallel manner.

At a personal level, by writing this book it is my desire, even if only in some small way, to have an impact over these next 15 years. Towards this end, the orientation was simply to capture my observations, experiences and thoughts, and pass them forward to others who hopefully will become actively engaged in addressing this challenge.

Ideally, others will build further upon what I've captured—extending, adjusting, and refining the dynamic, with their results achieving a broader and greater impact in the lives of those living in the rural areas of the LDCs.

...experiences and observations from the author

Chapter 1 provided an overview of the explosion of rural telephony in the U.S. during the early-to-mid 1990s. There were several contributing elements that made this possible:

1. On the legal and regulatory topic, the government was simply not in the way. There was no FCC;

2. There was a need in the rural communities that was unmet, as the monopolist, Bell, concentrated its attention on higher density, higher income, urban centers;

3. With the end of the patent on the phones, the technology became readily available in the marketplace and several sources provided small telecom solutions; and

4. Local small businesses and cooperatives stepped in to fill the gap, to the point that at one time there were over 6,000 small rural telcos operating in the U.S. These small telcos averaged less than 200 connections (customers), and for several decades they were not interconnected—the technology could not send an analog signal over long distances.

Today, 100 years later, this telephone story from the U.S. sounds like the current global Internet situation where a communication need exists, but it is unmet in many rural communities. There are 48 LDCs and many rural areas in the LCCs where the potential exists to expand affordable Internet. In fact, this need still exists in the rural areas of the U.S. and other advanced economies!

There is currently some movement in this space, including lower-cost mobile services (VANU, etc.) and Wi-Fi based Internet services (Mawingu in Kenya, AirJaldi in India, Microsoft's Rural Airband Initiatives in the U.S., etc.). Many of these are off power grid and solar powered.

In summary, affordable technology solutions are now available. And as a rule, Wi-Fi frequencies are typically unlicensed, creating a situation where there is minimal government oversight. The Wireless ISP (WISP) business models have been around for some time and are now moving into rural communities. There are also several emerging low-cost backhaul solutions becoming available, with more on the way. And there is demand.

The challenge now is to achieve global mass deployments by the thousands, perhaps tens-of-thousands, where local entrepreneurs focus on creating financially viable small Wireless ISP businesses to serve the bottom billion rural population in the LDCs and LCCs. Supporting such a global dynamic will have a profound impact, a relatively rapid impact, and a global-wide impact, that could serve as a catalyst for expanding socioeconomic impact across our planet.

This is perhaps the single most important area needing attention and much can be done now to meet this challenge. Ultimately the next generation of micro-satellites, small-cell 5G technologies, drones, and high-altitude balloons will play a critical role. Unfortunately, today there are but a few limited engagements that are active in this space.

We need an Internet explosion of Small Cell and WISPs to meet this global need—an explosion that is like the small voice network dynamics that took place in the rural communities in the U.S. 100 years ago. Back to the Future! Hopefully the international development community will soon place a high development priority on Digitization in LDCs that translates into making this an on-the-ground reality within the timeframe of the SDGs.

Acronyms

A4AI	Alliance for Affordable Internet
ADI	A4AI's Affordable Drivers Index
CDMA	Code-Division Multiple Access
CSTD	UN's Commission on Science and Technology for Development
DD	Digital Development
DIAL	Digital Inclusion Alliance
DtD	Development through Digitization
DFAT	Australia's Department of Foreign Affairs and Trade
DFID	Britain's Department for International Development
ECOSOC	United Nation's Economic and Social Council
FITEL	Peru's Fund for Investments in Telecommunications
GBI	USAID's Global Broadband and Innovation program
GITR	WEF's Global Information Technology Report
GCI	U.S. State Department's Global Connect Initiative
GSMA	Groupe Speciale Mobile Association
IAEG-SDGs	Inter-Agency Expert Group on SDG Indicators
ICT	Information and Communication Technology
ICT4D	Information and Communication Technology for Development
IED	U.S. White House's Internet For Economic Development
IDI	ITU's ICT Development Index

INGOs	International Non-Government Organizations
ISP	Internet Service Provider
ITDSC	Armenia's Information Technology Decision Support Committee
ITU	International Telecommunications Union
IXP	Internet Exchange Point
LCCs	Least Connected Countries
LDCs	Least Developed Countries
LEOS	Low Earth Orbiting Satellite
LI	USAID's Leland Initiative
LMI	USAID's Last Mile Initiative
LTE	Long-Term Evolution
M&E	Monitoring and Evaluation
MCC	Millennium Challenge Corporation
MCI	GSMA's Mobile Connectivity Index
MDGs	Millennium Development Goals
MNO	Mobile Network Operator
MVNO	Mobile Virtual Network Operator
MVNE	Mobile Virtual Network Enabler
NBP	National Broadband Plan
NRI	WEF's National Readiness Index
OPIC	Overseas Private Investment Corporation
PPT	USAID's Program Technology Transfer
PSTN	Public Switched Telephone Network

SDGs	Sustainable Development Goals
Sida	Sweden International Development Agency
SWOT	Strength Weakness Opportunity Threat
TDMA	Time-Division Multiple Access
TOPIC64	Vietnam's Training Online Programs and Incubation for Communities
UN	United Nations
UNCTAD	United Nation Conference on Trade And Development
UNESCO	United Nations Educational, Scientific and Cultural Organization
USAID	U.S. Agency for International Development
USAF	Universal Service and Access Fund
USF	Universal Service Fund
USO	Universal Service Obligation
USTTI	United States Telecommunications Training Institute
VTF	Vietnam Telecommunication Fund
WTDC	World Telecommunications Development Conference
WEF	World Economic Forum
WiMAX	Worldwide Interoperability for Microwave Access
WSIP	Wireless Internet Service Provider
WSIS	World Summit on the Information Society

www.ingramcontent.com/pod-product-compliance
Lightning Source LLC
Chambersburg PA
CBHW070240230526
45470CB00002B/463